The Bigfoot Field Journal

Scott Carpenter

Dedications

To my family

Travis, Miranda, Ryan, Brad

for their support and love through this adventure called "Bigfoot Research"

To my friends

Dave Paulides

for pushing me past my comfort zone and opening my eyes new possibilities

Richard and Mary Hucklebridge

for being the voice of reason, a sounding board, giving me honest feedback, and helping edit this journal

Chuck Prahl

for always challenging me and keeping me grounded

Without your love and support none of this would have been possible

Contents

Forward

There are many people that call themselves a "Bigfoot Researcher," few actually earned the title. In the dozens of "researchers" I've met, the most competent, most innovative and the most persistent would be Scott Carpenter. Scott and I first met when we trekked into the Great Smoky Mountain National Park and he invited me into his study area. Scott is a very humble person with a wealth of knowledge. The only Bigfoot sighting I've ever experienced occurred with Scott in an area where he's had continued interaction and collected dozens of DNA samples. I know for a fact the sighting would never have happened if Scott hadn't explained where and how to look in order to make the observation. No, this wasn't a Patterson-Gimlin type sighting, I saw the biped watching us from a very well camouflaged position, the same position where they are probably watching us on trails on a daily basis. It was a very surreal experience that I will never forget.

While every other "researcher" was placing trail cameras in remote areas attempting to catch the biped on film, Scott realized that a new innovative approach was needed if we wanted to a Bigfoot in a photo. Scott started to place small cameras on the trails looking behind him as he walked into the mountains. This unique approach to the issue has gotten Scott and NABS team members a very unique look at Bigfoot that few have ever seen. This has also proven that the biped is following us as we do our work and are probably watching the entire time were in the woods.

One of the most surprising incidents to ever occur to me in the mountains also occurred in the Great Smoky Mountains National Park. Scott and I were hiking into a research area when we were stopped by what sounded like teenagers yelling at each other from a very, very thickly vegetated area deep in the wilderness. We stopped and listened with great amazement as this continued for a few minutes. Scott had his video recorder rolling and captured the sounds. The sounds were forwarded to linguistic expert Scott Nelson who confirmed they were Bigfoot language!! Again, if Scott hadn't had the video rolling, nothing would've been recorded and the memory would not have been preserved.

Inside the cover of this book, you are going to privy to some of the most unique photos and research ever captured on Bigfoot Scott is one of the nicest and most credible individuals I have met in this research arena. Trust that Scott is presenting quality research that few have ever had the privilege of viewing. NABS is very lucky to have him on our team, you are fortunate he is willing to share.

David Paulides
Executive Director North American Bigfoot Search
Author of – The Hoopa Project, Tribal Bigfoot, and Missing 411
October 2012

Preface

First of all thank you for purchasing my book. I trust you will find it an interesting and enlightening read. This is my first endeavor into the literary world. Actually this is more of a journal than a book as the title implies.

This journal will take you through my experiences during the two years of my new found career as an amateur Bigfoot researcher. I think it would be prudent at this point to let the reader know that this book or journal does not try and prove the existence of Bigfoot. It is a collection of my experiences and pictures that I gathered while out researching the Bigfoot phenomenon. I come at this from the point of view that Bigfoot does exist and is a real biological entity. That he eats, breaths, and lives in the forest. It is not of matter of "I believe" Bigfoot exist, to be frank I know he exist. The question "do you believe in Bigfoot?" is like asking me "do you believe in the sky and the sun above".

Unlike many books on the subject, you will find many still capture photographs from the videos I have recorded in the field. I learned the hard way that you should have your video camera recording at all times when you are out in the woods. The average video camera takes approximately ten seconds to power up and be ready to take video. The average Bigfoot encounter last about five or six seconds! I run the video camera from the time I step out of the truck to the time I get back in the truck to leave a research area. I have hundreds of hours of video from my expeditions, and it takes an enormous amount of time to review these videos, but that is the nature of this endeavor. Patience and persistence is the key. So you will see photographs of what I consider to be Bigfoot. You will see photographs of many different Bigfoot and the evidence they have left behind.

I realize some will look at these photographs and see nothing or exclaim that they are "hoaxes" or "anomalies". That is also the nature of this type of research. Everyone has their opinions on the subject. The debate still rages over the Patterson/Gimlin film, so I am aware that my photographs will not be accepted by all and that is fine. I am presenting my experiences and opinions not so called "proof".

I would like to point out that every photograph I have put in this journal has had a thorough follow up. If I find what I think is a Bigfoot on a video I will return to the area and take follow up video, look for trace evidence like footprints, broken limbs, disturbed vegetation, hair, etc. I strive to make sure that the image I think is a Bigfoot is not an anomaly caused by shadows and clumps of vegetation. I actually try to "debunk" my own work. Every image that I think is a Bigfoot has been thoroughly researched and trace evidence has been found in the area where the video was taken. I will often compare before and after photos of the area and will show the trace evidence. If I have a photograph that is speculative then I will tell you upfront.

It my sincerest hope that you will enjoy this journal of my experiences with the people known as "Bigfoot". I hope you will gain a better understanding of Bigfoot and also what it is like to pursue the most elusive race of people on the planet. What I often refer to as the "Lost Tribe".

Introduction - How it all started- My first experience

This experience has stuck with me through the years and has been one of those occurrences I could not get out of my head. I was eight years in 1971 when this event occurred. I lived in rural north Knox County, Tennessee at the end of a dead end gravel road. We were surrounded by small cattle and tobacco farms, with mixed in wood lots. It was the summer time and I had stayed up late with my mother watching television. My father was working the night shift that evening, and my younger siblings were already in bed. I remember the news broadcast had ended and it was around 11:30 PM. My mother stood up and told me she was going to take the garbage out to the drum before we went to bed. Back then everyone had a 50 gallon metal drum with a heavy lid to store their garbage. The back door was opened, and my mother went off the porch leaving me on the couch. The garbage can was located at the back of our lot approximately 75 to100 feet from the house. Immediately after my mother walked off the porch these two little "monkey" looking creatures ran on to the porch, put their faces against the glass door, and looked in. I was scared to death at first, but when I made eye contact with them they both smiled at me and my fear subsided. The best way I can describe their appearance is "monkey" looking. They resembled a chimpanzee, but the mouth was human, the skin on their faces was gray, their eyes were very black, and their noses were little human looking noses with wide nostrils. I heard the lid to garbage drum slam down and they jerked their heads around, looked in my mother's direction, and then ran off the back porch. I did notice when they ran off the porch they left their arms above their heads and ran like a chimpanzee runs when it walks on two feet using its arms to balance it.

When my mother arrived back at the house she entered the door complaining "there must be a skunk in the field, it smells horrible outside." I asked her if she saw the "monkeys" that were on the back porch and she gave me a funny look. I insisted a couple little monkeys were looking at me through the screen door. She quickly dismissed it and blamed the incident on our neighbor's children playing a prank on me. I knew it was not my neighbors, but I just let it go and went to bed.

I have tried to remember more details about their faces but all I can clearly remember is how dark their eyes were and how they smiled at me. They had straight white teeth and regular lips like a person. One was approximately four and a half feet tall and the other one about a half a head smaller. I would approximate four feet. I remember that the arms were long for their body size and how they ran off the porch resembling a running chimpanzee. I had nightmares for many years over this experience and until recently never made the connection in my mind that these could have been infant Bigfoot. After reading and studying about Bigfoot and Bigfoot encounters I made peace with this memory coming to the conclusion that they were Bigfoot infants.

Location Designations

I have decided not to divulge the locations where I perform my research in order to protect both the Bigfoot and the people involved. I hope the reader will understand that in any discipline there are those that seek to take advantage of and exploit others for monetary gain. The field of Bigfoot Research is no different. Unfortunately there are those that would use the information is this book to exploit the Bigfoot and do irreparable harm to them. It is because of this I will designate the location that is on public land "Research Area A" and the location that is in the National Park as "Research Area B".

In the right place at the right time

Date: June 27th, 2009
Location: Off the shore of Research Area A

I had been watching amateur Bigfoot Researchers on YouTube for a couple months. This rekindled my interest in Bigfoot and I had begun doing some reading on the subject. I had read several accounts where a Bigfoot reportedly made what is known as a "whoop call". I went to a couple websites that had an alleged recording of an actual Bigfoot making the "whoop call". I listened to the calls and thought to myself next time I go fishing and I am off the shore of a remote area I might try imitating this call to see what would happen. I did not really think I would ever get a response, but what the heck, never hurts to try.

The Saturday of June 27th, 2009 I was bass fishing a point of land on Research Area A. Before making my first cast I made five or six "whoop calls", doing my best to imitate the calls I had heard of the alleged Bigfoot recordings. To be honest the calls were not very good, but by my last one I think I had it sounding pretty good. After I finished I felt a little silly and hoped there was no one else on the lake that heard me. I reached down, picked up my fishing rod, and began to cast a top water lure to the point. I made maybe three of four cast when something came crashing loudly through the woods toward the water. I was shocked! I froze in mid-retrieve and began looking into the woods. I could not believe what I was hearing. It sounded like something very large running through the heavy underbrush breaking limbs and making a great deal of noise. It was making too much noise to be a small forest animal or even a deer. There were no horse trails that came to the end of this point. The entire shoreline was grown up in thick underbrush.

I focused my eyes on the shoreline trying to detect anything moving. My eyes caught movement back up in the woods a couple times, but it was just too thick to see anything clearly. I did hear what sounded like bipedal walking up and down the shore line like it was looking for something. I also heard what sounded like a low, deep, "hummmph". I heard it three or four times very clearly. Whatever it was went back and forth for approximately two minutes, stopped, made the "hummmph" sound and then walked away slowly back up into the woods. My heart was beating so hard I could hear it in my ears. I could not believe what had just happen. Surely this was not a coincidence. What animal in this area would respond to a "whoop call" and tear through the woods like that? This shook me up and it took me a few minutes to regain my composure.

I kept asking myself as I resumed fishing, "Was that a Bigfoot?", "No, It couldn't be, could it?", "What else was it then". I spent the rest of the day thinking about what had happen and what it could be. I was not sure what it was but one thing I knew, I had to go in there and look around. I had to take my camera in there and see what made this noise and see if I could find any evidence of what made it. This incident began my journey into the Bigfoot Phenomenon as an Amateur Researcher.

Research Area A is a 1,200 acre public area. It is located in East Tennessee. The area is a multipurpose area that allows hunting, hiking, and horseback riding. There are no motorized vehicles allowed. It is walk-in or horse back only. The fields are leased to local farmers who grow corn, soybean, and winter wheat.

YouTube "Big-Footer"

After the "whooping" experience I began to search YouTube for more Bigfoot videos. I noticed two things. One, there is a dedicated group of amateurs that would venture out in the woods and then post snippets of their trips on YouTube. Two, most of the videos that claimed to show a Bigfoot was blurry, out of focus, or the subject was so far away it could not be identified. I thought to myself "why are the researchers not enhancing the videos?" I looked at many that had potential, but no one was enlarging and trying to clean up the images. Many of these videos were of known animals, artifact, shadows, clumps of leaves, etc. I often wondered if some of these videos could actually be images of a Bigfoot. The only way to find out was to begin enhancing them.

I started with the videos that were being posted by amateur researchers. This was the most logical approach since they were actually looking for and researching Bigfoot. Many had posted images of possible footprints, limb breaks, unusual sounds, and possible images of Bigfoot. I am an Information System professional by trade. I started my career in the Air Force as a main frame computer operator in the early to mid-1980s. Once I returned to civilian life it was the natural career choice. I have over 25 years 'of experience in the field so learning about video and picture enhancing software and techniques came easy. Raising a family of four children and caring for a disabled wife left little extra cash for the purchase of enhancement software. So I began researching and trying out "freeware" and "shareware"programs. I also had an old version of Adobe Photo Shop that had come with a PC I had purchased years earlier. Through experimentation and downloading trail versions of high end video enhancement software I soon learned about and found shareware video enhancement software that performed as well as the higher end programs. There were some limitations, but considering my budget this would do.

I soon began enhancing and posting my "finds". In the beginning I was seeing Bigfoot in every clip, but I learned what was artifact and what was not with some practice. I also discovered that the knack I have had since childhood to spot wildlife was helping me find some interesting things in the videos I was reviewing. I am no expert and do not claim to have some "special" talent, just an eye for spotting shapes and slight movement. I use this skill often in deer hunting. You do not look for the entire animal, but pieces of the animal. You look for an eye, ear, leg, tail, etc. I began to retrain my eyes to look for parts of a humanoid. Round head, shoulders, eyes, nose, leg, hand, fingers and soon I became pretty good at spotting what others were missing. Of course my critics claim what I see is "shadows, leaves blowing in the wind, and wishful thinking". So what I claim to see and enhance is always open to interpretation and it should be.

I began to think that I could do this myself. I have a video camera and some woodsman skills, why not go out and look for Bigfoot? Sitting in front of a computer screen trying to enhance what other people were finding was not getting me any closer to finding out what had responded to my whoop calls. It would also help me with the enhancement work to know the challenges of trying to document a research outing with a video camera.

I started small with a couple trips to a hiking/horse trail in the nearby National Park to get my feet wet. After this I was ready to hit the area where I had my experience. And so now I had officially joined the ranks of the Amateur Bigfoot Researchers of YouTube.

Chapter 1 - First Expedition

Date: July 12th, 2009
Location: Research Area A

This is my first "official" research outing. While my wife and kids are enjoying the sun and the lake on the shore line, I am dressed in my "OD Green" pants, light weight camouflage shirt, a "Boonie" hat, and boots. I have my camera and back pack on ready to go discover the "beast" that made all that noise just a few short weeks ago. Little did I realize what I was getting myself into.

It was very hot. The temperature was in the mid 80's as I followed the horse trail into the woods. I followed the main trail until it came to a T. I was looking for the location where I had heard the loud noise while fishing, so I turned right and went down the very thick overgrown horse trail. I did not know if I had traveled far enough into the woods yet, but I thought I would head down this horse trail to see if it led to the end of the point.

I walked down the overgrown trail until it dead ended into the lake. There was an old road bed covered by about a foot of water. On the other side of this road bed was a small island. It was obvious I had turned down the wrong trail and as I looked to my left I could now see the point where I had heard all the noise. So I turned around and started back up the trail.

Walking back up the horse trail I noticed what looked like a faint game trail leading off to my left. I looked to my left and noticed what appeared to be vines and sticks intertwined. I decided to investigate this odd looking formation. When I walked up to it I noticed it had the look of something that was not naturally made. Vines, tree limbs, and blow downs were all part of this "formation". Some limbs and vines looked intertwined as if they were wrapped together. I looked in it and realized this could be a "shelter". I looked around it thoroughly and noticed a faint urine smell.

The "Shelter" from the front side

After documenting the "Shelter" thoroughly I made my way back out to the horse trail and started working my way around to the point where I had the encounter after making the whoops. I found a crossing trail about a quarter mile up the trail and took it to the beginning of the point. The trail split again and I turned right to go toward the point. The trail was grown over and had dead falls blocking it, so I went off trail and began working my way through some heavy under growth as I made my way into the woods.

I work my way down through the thick foliage and find a faint trail with high head clearance. A normal deer or game trail will have a clearance of approximately three to four feet. This trail had a clearance of seven to ten feet with small limbs broken off a couple feet above my head. Making my way closer to the end of the point where I had the response to the whoop call I began to find fresh broken saplings and tree limbs.

Along with the numerous sapling breaks I noticed the ground was literally covered in Black Gum Fruit. The food sources in this area were definitely abundant.

I had made my way about three quarters the way down the point when my cell phone rang. It was my wife who was ready to go and get out of the oppressive heat and humidity. This expedition would end before I could investigate the area thoroughly. Frankly I do not blame my wife, it was a very hot day and I was also glad to get out of the heat.

Chapter 2 - Infrasound and First Encounter with the Bigfoot Clan

Date: August 1st, 2009
Location: Research Area A

This would be my first "all day" outing. I was not prepared for what was going to happen to me. I would experience things that seasoned researchers seldom have happen to them.

I arrived at 8:30 am and started down the trail. My first discovery is a box turtle creeping across the trail. This would be the first of many turtles I would find this year. I would also find many empty turtle shells. I continue down the trail and find a side trail cutting off to my right and decide to go down it to see where it leads. The trail leads down to the lake shore. The area is full of thick vegetation and as I walk down the faint trail I find a curious looking structure. It appeared that two cedar trees had fallen and formed what appeared to be a hide or blind. It was just a few feet from the water's edge but gave a good view of the horse trail while providing excellent concealment.

These two trees form the "Hide" or "Blind"

This is an intriguing structure and I start looking around it. The first odd thing I noticed was a couple broken limbs that look like they were placed on the side of the structure.

I look into the center of the blind to find the vegetation compressed. I also notice fresh tender plants, flower blooms, red bud pods, sheep sorrel, and honey suckle scattered in the floor of the blind. It appears that something was laying here and feeding recently. The blooms and tender plants had not begun to wilt. Given the heat and fresh condition of the blooms whatever was in this blind had just left the area that morning if not moments before I arrived! I work my away around the back side of the blind and get down on my hands and knees and crawl in to look around. Once I crawl into the blind I found that some of the ground cover had been pushed away and left exposed dirt. One of the clean areas has the shape of a lower arm or shoulder where a Bigfoot could have possibly propped itself up by its arm.

It is hard to see in the below photograph but a hair pattern was present in the exposed dirt. The yellow squares show tender leaves and honey suckle blooms.

Impression in the ground and scattered fresh blooms and leaves

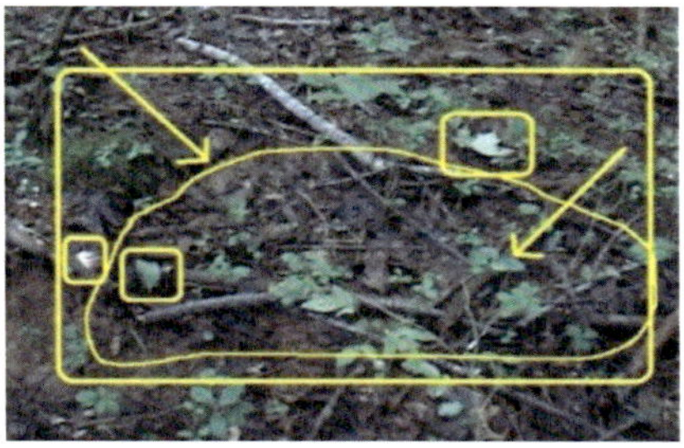

The floor is also littered with flower blooms, honey suckle, sheep sorrel, red bud seed pods, and plucked tender plants that are not present in the immediate area. These items must have been carried to into the blind by something.

After documenting the blind using my video camera I just sit down for a minute and notice how well I am concealed. I also noticed I had a good view of the horse trail and could watch people and horses pass without being discovered.

I noticed other characteristics that made this an ideal location, the lake to the rear so nothing could come up from behind the occupant, two avenues of escape into heavy brush, the cool breeze coming off the lake, and the easy access to water while staying concealed. This would be a very nice setup to sit and watch people and horses.

I complete my investigation of the blind and make my way down a trail that follows the shore line. I go approximately one hundred feet and I find a limb that has been snapped off, then placed in the "Y" on the same tree. I wonder how this limb broke then went up three feet, then came to rest in the "Y" on the same tree? The broken end with its own "Y" was placed up into the tree with the small straight end of the limb pointing up. It is difficult to come up with an explanation other than the obvious. The limb was broken off and placed there. Could this be a trail or territory marker?

When I take the limb out of the notch and match it up with the broken end still attached to the tree it matches the break perfectly. How did it break off, travel up several feet and lodge in the notch above it?

I continue down the trail and decide to turn off the camera for a few minutes and just listen and watch. This would prove to be a mistake I would not repeat. After turning off the camera I stand very still and just listen, then without warning I hear a series of very loud splashes in a mud hole about fifty feet from where I am standing. I would estimate I heard at least three or four splashes. Of course with the camera off we will never know how many. I turn the camera back on and begin looking at the horse trail that I think is at least thirty feet away from me. I later find out I was much closer to the trail, only about fifteen feet away.

When I first hear the splashes I thought it was a horse and its rider coming down the trail. Not wanting to spook the horse I just stood still and waited for them to pass. I waited five minutes, then ten, and no horse and rider pass. There was no way I missed them. I began to get nervous. Did I hear something else? Was this a Bigfoot? About the time I was thinking these thoughts I smelt something I had never smelt before. The only way I can describe this smell is to say it was like a "wet dog had been sprayed by a skunk". It was a sickening sweet smell that almost made me gag.

I stood still for a few minutes wondering if a Bigfoot was coming toward me or going away from me. I finally gathered my nerve and decided to go see what had splashed through the mud hole. I worked my way toward the horse trail and came upon it very quickly and realized I was only about fifteen feet away. If a horse and rider had traveled past me on the trail I would have noticed them.

I worked my way back to the mud hole slowly listening and watching. The brush was very thick on both sides of the trail. Visibility was limited, anything could be lurking just a few feet away and I would never see it.

I made my way to the first wide area in the mud hole and started looking for footprints. The mud hole was so full of horse hoof prints and water that anything could run through the area without leaving a footprint. I did however find a couple of "possibilities".

One possible footprint appeared to be made by something moving across the trail not down the trail with the other horse traffic. It is not conclusive and proves nothing, but it is another one of the oddities that would reoccur many times.

Infrasound Experience

I keep moving down the trail looking for footprints when I begin to feel heaviness in my chest. I then start finding it hard to breathe as I begin to get a sense of panic. I begin to feel oppressed. I do not hear an audible voice but I get the intense feeling I need to leave and get out of the area. I was able to recognize what was happening as I began to get emotional and my eyes were tearing up. I had been doing research and had read about Bigfoot's ability to make people feel panicked. So I stood my ground and spoke out loud to whatever was making me feel this way. I said, "You are not going to make me leave, not this time! You let horses pass by here all the time so you can just get over it!"

This seemed to work and the feelings begin to subside. I am sure you are wondering what is "Infrasound"? Infrasound is a low frequency sound wave that the human ear cannot detect. Depending on the frequency infrasound can have many effects on the human body. It can cause the eyeballs to vibrate, nausea, disorientation, memory loss, confusion, anxiety, and paranoia. Certain frequencies can render a person unconscious or even cause death. Large mammals such as Tigers use it to confuse and disable prey. It is theorized the Bigfoot use infrasound to communicate, avoid human contact, and disable prey. (Source: http://www.oregonbigfoot.com/blog/bigfoot/nvcode-part-eight-infrasound/).
All I know is I experienced something tangible that had a profound effect on me. It had both a physical and emotional impact on me and again later in this day I would experience it again.

The Stick Shelter Project Begins

I continue down the trail and head back toward the stick structure I had found on July 12th. I had walked a few feet down the trail when I heard something to my right in the thick brush. It sounded like footfalls of something struggling to keep pace with me through the thick underbrush. The sound was too my right rear and I heard the noise until I almost reached the area of the stick structure.

Once I arrived at the stick structure I looked around to see if anything had changed since I had last visited. I found a large flat rock that was not present during my last visit. I wondered where it came from. The rock is approximately 12 inches long and 4 inches wide.

I had brought some rocks with me to leave in the area. When I saw the flat rock the idea hit me to place the flat rock under the shelter and put the round rocks I brought on the top of the flat rock.

Below is a close up of the three rocks that I left on August 1st, 2009. Take note of the color of the middle rock and the shape of the smallest one. Also note the bottle I marked to the left. The rocks and bottle will take a "trip" in the near future.

I had set what I called my "Rock Trap" to see if anything, hopefully a Bigfoot, would move or take them. I left this area and headed down the horse trail to explore more of the area.

I see a "Blob" and Infrasound Confuses Me - I am surrounded by a Bigfoot Family

I make my way down the narrow horse trail that is almost completely overgrown. It is very hot and humid. The noise made by the cicadas is now extreme. The volume is greater now than it was earlier in the morning. My plan is to work my way over to where I had the response to the "whoop" calls a few weeks ago. On my first expedition I had to turn back before I made it out to the area where I heard the noises and limbs breaking.

A few hundred yards up the trail I find a small side trail that leads down to the lake shore. I do not know why, but I decided to walk down the trail and check it out. Little did I know that this decision would put me within a few feet of a large Bigfoot and several members his clan.

I walk to the lake and find an area of brush and small trees with the trail splitting to the left and right of this area. I decide to go the right side first and make my way toward the lake. Unknown to me at the time, at least two Bigfoot have just ducked into the thick foliage on the left. They were moving out of the area of brush and small trees. An adolescent Bigfoot was trapped and is still hiding in this center section. I move to my right and begin investigating the area. I find what appears to be a bedding area in the grass. There is a limb twisted, broken, and pulled down in front of the bed as if to provide more concealment for whatever had been laying here. I kneel down and look around and I discover two matted wads of course hair. I decide to keep one of the wads of hair and put it in my pocket and drop the other one on the ground. I also notice from the location of this bed you can look out on the lake and watch the boats go by in the shade and be totally concealed.

I turn around to work my way back up the trail. As I am turning I capture a few frames in the area formed by the "Y" of the splitting trails. I was not aware of it of the time when I was filming but when I reviewed the video I noticed something very suspicious. Though it is not crystal clear it is a small Bigfoot.

The infant Bigfoot watching me

This still capture shows the left eye, cheek, nose, mouth and forehead

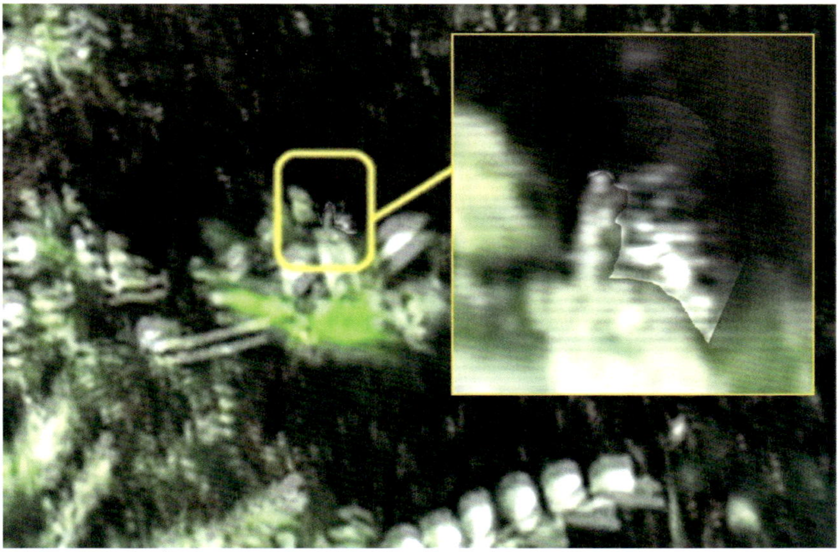

I make my way around the center area of the "Y" and head down the left trail. I go down to the water's edge and look around for just a few seconds. Though I do not realize it I am only a few feet away from at least one Bigfoot. I turn and go back up the trail a few feet looking into the foliage when something begins running through the thick underbrush. I thought at first it might be a deer but it stopped after approximately one hundred feet. I think it was what I call the "Rabbit". It is a common strategy for one Bigfoot in a group to take off running in order to draw your attention. This will allow the others that are hiding nearby to sneak off in the opposite direction while you are watching the other Bigfoot exit the area.

This ploy was unsuccessful since I was aware of this tactic. I froze and pointed my video camera into the brush. I concentrated for a few minutes on the brush where I heard the sound then moved back down toward the lake a few feet. I was looking into the woods when I noticed a large dark "blob". I even commented out loud "I see a blob". It was like something very large had literally entwined itself in the middle of a cedar tree. I could not focus my eyes on it because the figure was distorted and blurry. This is the point where things get weird to say the least. I think I was exposed to infrasound again, but this was different than the first time. I had every intention of stepping forward and trying to see what the blurry blob actually was when my cell phone went off and broke my concentration. When this happen, without explanation, I turn and walk off muttering to myself about who was calling me. I do not have an explanation of why I just walked off. I also lost my memory of the events that occurred over the previous 10 minutes. I was confused and just walked off. I did not remember what happen to me until I returned home and was emptying my pockets and pulled out the hair. The moment I touched the hair and pulled it out of my pocket the memories of what happen came back to me. I am convinced this was the effect of infrasound. The cell phone going off broke my concentration and I was disoriented and confused by the infrasound being emitted by the Bigfoot.

After remembering what had happened I hurriedly reviewed the video I had taken that day and found that something was in the cedar tree and it was recognizable. During my turn to walk back up the hill I captured several frames of a gray face looking out of the cedar tree. It was well camouflaged and only comes into focus for a few frames but in my opinion there is a very recognizable face. The face is at least ten feet off the ground. This was a large Bigfoot!

I was disappointed that I not able to capture more video of the Bigfoot. The still captures I have are from a moving camera and are not as clear and focused as I would like them to be. I was able to capture roughly thirty frames of the Bigfoot. This would equate to only one second of video at thirty frames a second. In the world of Bigfoot research one second is a long time. The average Bigfoot encounter only last approximately five seconds. Even though the quality was not what I hoped for at least I had some footage that was usable and showed reasonable detail.

This is the frame that caught my attention. The "face" is gray with a black nose. The height of the Bigfoot is at least ten feet. It is standing approximately fifteen feet away from me.

This is a common characteristic reported by eye witnesses across the country. The interesting thing is the black patch on the nose. I did research on the sightings in my area and came across a BFRO report where the Bigfoot had a "Black Nose". The sighting was 20 years prior to this one, but I found it interesting that the location was only 25 miles away from where I had my sighting. (BFRO Report# 10771)

On the previous page, the photograph on the left is a blow up of the face and the drawing on the right is my interpretation of what the Bigfoot looks like based on the photographs. Below is a close up detail of the Bigfoot.

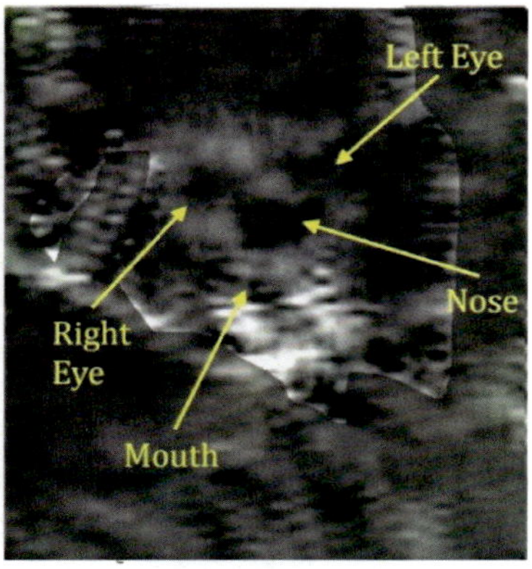

Only 24 four frames, less than one second of video. A brief glimpse of what I think is a Bigfoot. Not only a Bigfoot, but what I think is the patriarch of the clan that inhabits this area. He was protecting his own by using infrasound on me. He did not have to result to violence which he could have and dispatched me with no problem. Instead he used stealth and confusion to defuse the situation. In the pictures below I think I have found one of the sub-adults looking out at me from the brush. This guy was even closer. I my opinion I see a set of eyes reflecting the sun.

Just under the large Bigfoot in the mid-far-left of the frame is the "Little-One"

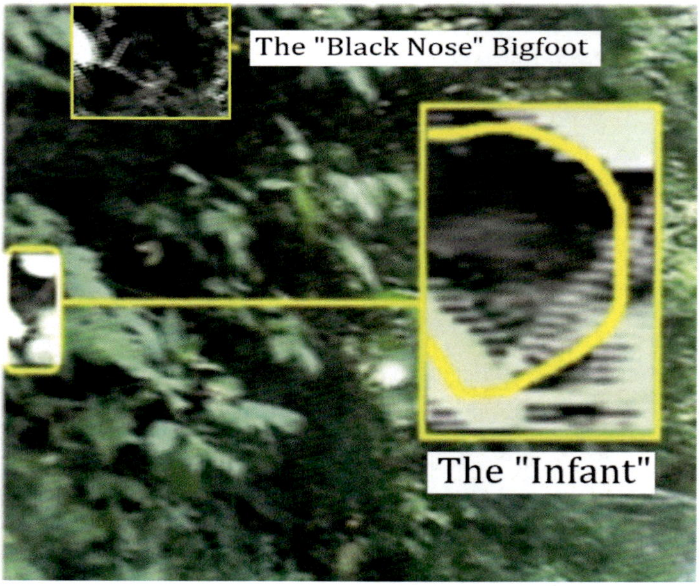

After getting a message on my cell phone I wander off back up the trail leaving the area. When I get back to the main trail I begin to smell the "Bigfoot Smell" again. It is very strong, but I am confused about the smell because the effects of the infrasound have disoriented me and caused me to forget what just happen.

I continue walking the trail and cut across to the point of land where I had the reaction to the whoop calls a few weeks earlier. I wanted to make it all the way to the end of the point and look for any evidence of what made the entire ruckus just a few weeks ago.

I worked my way down the point of land. The vegetation was now very thick. I found an old grown over road bed and followed it toward the end of the point. Walking along I noticed some sign that something large had walked down this road bed. I am not an expert tracker, but I do know how to track relatively well. There is no way an animal that weighs between 400 and 800 pounds can move through the forest without leaving some sign. Bigfoot are experts at camouflage and stealth, but they cannot fly to my knowledge and they do disturb the forest floor. They can be tracked if you know what to look for. I found a few large broken limbs in the trail that were at least two inches in diameter. It would take a great deal of weight to break them. I was at least a quarter mile off the nearest horse trail and there were no hoof prints or other signs of horses.

I found some old limb beaks and some faint sign but two months had passed since the whooping incident so there was no obvious sign of what made all the noise. I worked my back out of the area and went home. I stated before in this chapter that once I returned home my memories of the event came back to me when I reached in my pocket and found the hair. I down loaded my video camera and begin watching the video. I could not believe what I saw. There was this face in the pine tree where I had seen the blob. I was confused and could not understand why I just walked away. I was extremely upset with myself for just walking off. Why did I not step forward? I saw it, even mentioned "I see a blob" on the video then inexplicably just walk off. I would later conclude that I was influenced by infrasound emitted by the Bigfoot.

I showed my wife the matted hair I had found. I was not sure it was hair, but my wife was very positive it was hair. It was course and matted together, much like a dog that has not been groomed will get. I would eventually send this hair off to be tested for DNA. The lab reported back that the follicles were too small to extract a viable DNA sample.

After seeing the video and pondering the hair I had found I knew I had to go back the next day. I had to see if the face was just artifact or shadows and light. I had also left a second batch of hair in the bedding area and I wanted to go back and get it. I knew I had to return, I had to answer all the questions that were nagging me about the video and the matted hair I had found.

Chapter 3 - Follow-Up One day after Sighting

Date: August 2nd, 2009
Location: Research Area A

I had a hard time sleeping that night wondering about what I had seen. Looking at the video and seeing the face looking out at me. I knew I had to go back and look for evidence, hair, footprints, and get the swatch of hair I had left behind.

I woke up the next morning and prepared to go back out to the area of the sighting. My two youngest children wanted to go so we all loaded up in the truck and headed out. It had rained during the night and I was afraid that any footprints that may have been left behind would be washed away. Once we arrived we quickly made our way to the area of the sighting. When I found the sighting location it became obvious that what I had seen and captured on video was not there. I was very frustrated with myself. Where the face and the dark body mass had been the day before there was now open daylight coming through the brush and cedar tree. I took my video camera and tried filming several different angles so I could duplicate my position from the day before. The more I looked and filmed the more I realized that there was something in the cedar tree the day before, something with a face and something very big.

The photograph on the left shows the area where the Bigfoot was hiding the previous day. The face and dark mass are not there.

August 2nd, 2009

August 1st, 2009

A closer look at the same location again shows light passing through the area where the face was just the day before.

After taking video of the area I go into the brush and look up into the cedar tree for any sign, hair, tracks, etc. Once I started into the brush I began to find dozens of small broken limbs. The disturbed foliage in the brush showed evidence that something had moved through it. Dozens of small limbs were broken inside of the cedar tree. The broken limbs started at approximately shoulder high and went up in to the cedar tree at least twelve feet high. It appeared that the Bigfoot was hugging the tree, wrapping its entire body inside the tree in order to be as camouflaged as possible

I also found a partial footprint that I think came from one of the smaller Bigfoot hiding in the brush as I was being bombarded by infrasound. The boys and I were getting ready to leave the area when I noticed the imprint of toes in the wet ground right in front of the area where I captured the video of this little guy. The prints were leading out of the brush where the big one and little one was concealed.

After documenting the footprint the boys and I left the area. I was very disappointed that I had been affected by the infrasound. I wished I had stepped toward the brush and investigated the "blob" more thoroughly. I could have had a video of several Bigfoot fleeing as I entered the brush or I could have been torn limb from limb. In a good way we will never know.

Chapter 4 - The Rock Exchange Begins

Date: August 8th, 2009
Location: Researcher Area A

The following week my daughter and her boyfriend ask to go see where I had my encounters the week before. I was glad to take them and hoped their presence might invoke a response from any of the Bigfoot in the area. We arrived and made our way to where the "Stick Structure" was located. I showed them the area and explained that I had smelt a very strong urine smell the week before. I also explained that the flat rock was placed in the shelter the week before so I left three rocks on it as a "gift". I noticed that the rocks had been moved a little but it was not until I returned home and compared the rocks from the week before that I noticed they were different. In fact two of the rocks I had left on August 1st, 2009 had been taken and two different ones left in their place! I was stunned at first, but given the previous week's events and sighting it would not be out of place for something like this to happen.

The above photograph shows how I had left the rocks on August 1st, 2009. Note the clear glass bottle to the left of the rocks.

The photograph below shows how I found the rocks when I returned on August 8th, 2009.

Notice the dark rock is gone, replaced with a stone of similar size but lighter in color. Also the small rock with the notch has been replaced with a rock of similar size but different shape. Also notice that the clear bottle that was to the left of the rock is missing.

The dark rock and the rock with the notch have been replaced with other rocks and the bottle is missing. This is what I would consider an exchange with a Bigfoot. My rocks were accepted and I was left others. Why the bottle was taken I do not know. This is the beginning of a ritual between me and an unseen Bigfoot that will go on for months.

Below is an enlargement of the rocks before and after. Note the dark rock and the small rock with the notch in it. On August 8th, 2009 when I return these two rocks are gone and been replaced with rocks of similar size but obviously different. I have read of many accounts of people exchanging gifts with a Bigfoot but I was stunned that it had actually happen to me. In two weeks' time I had experienced infrasound, found hair, found footprints, and stood within a few feet of an 8-10 foot tall Bigfoot and other members of his clan.

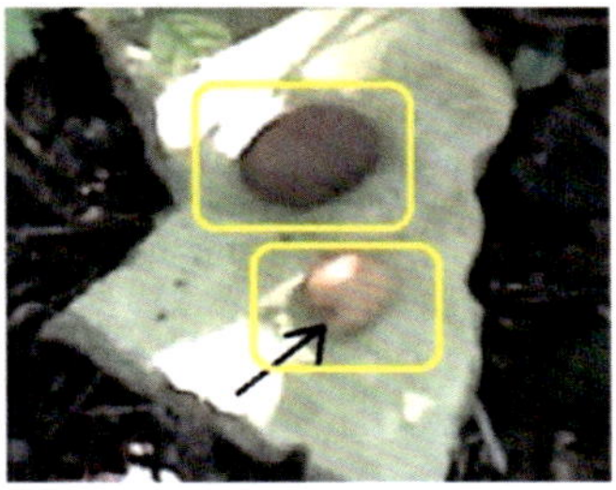

Rocks left on 08/01/2009

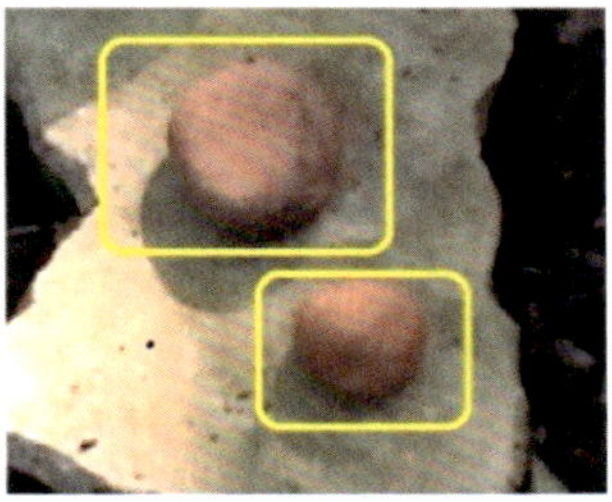

Rocks given to me on 08/8/09

The other interesting thing to me was why did it take the bottle? Had stumbled onto some short of shelter? Where they collecting trinkets that were important to them?

We looked around the shelter and I arranged the rocks again in a nice straight row before we left the shelter area. On the way back out the trail we came back to a large mud hole that was approximately fifty yards down the overgrown horse trail. Going around the edge I noticed a single impression in the mud that was in the shape of a large foot. It was at least fourteen inches long. I did not have a measuring tape with me and the possible print was in the middle of a still wet and sticky mud hole so I took a few video shots and we continued up the trail.

After documenting the possible print in the mud hole we walk down the trail to the area of the August 1st, 2009 sighting. I wanted to show them where the sighting happened and do some more documenting of the area. My daughter's friend's height and weight are comparable to the Bigfoot that was squatting down and hiding in the bedding area near where I found the hair. I wanted him to sit in the same place so I could do some comparison photos. I also wanted to look for any trace evidence where the big one was standing in the cedar tree.

When we arrived at the location I showed them were I found the hair and pointed out where the little one was sitting. My daughter's friend went around to the other side of the bedding area and made his way to where the Bigfoot was sitting back on August 1st. When he located the area where the Bigfoot was sitting he found the broken vegetation that had been used to hide the face of the Bigfoot.

Below is a side by side comparison showing the Bigfoot on August 1st and the broken vegetation a week later.

08/01/2010 - Little One Hiding Face

08/08/2010 broken foliage used to hide face

After looking at the bedding area we moved over to the area where the Bigfoot I call "Black Nose" was hiding in the cedar tree along with at least one or two smaller Bigfoot. We look for things I missed the week before and begin to document the new trace evidence that is found. I first take a few shots of the area where the face was visible to verify that it was not an "artifact" caused by shadows or the digital camera. I also looked at the vegetation in the front of the area where they were hiding. I began to find broken vegetation and poison oak vines pulled off of a tree. It became apparent as I investigated that the Bigfoot went into the brush from the front when they heard me coming down the trail. I must have surprised them or for some reason they could not leave the area. I often wonder was one of them sick or wounded? Did they have young that could not escape without being exposed so they chose to hide instead? I find it hard to believe I really surprised them; did they just want to get a closer look at me? Had my refusal to leave the week before when bombarded by infrasound raised their curiosity? Did they want to see who was trading the rocks with them? Whatever the reason, they went into the foliage and left sign of their presence behind.

We finish up our investigation of the area and head home without incident.

Chapter 5 - A New "Gift"

Date: August 16th, 2009
Location: Research Area A

I rose early in anticipation of what I would find at the shelter area. Had the rocks been moved around? Have they been taken and replaced with others? I made the twenty five minute drive to the research area and then made my way to the shelter. When I arrived the rocks had only been moved a small bit. This was more than likely a small animal. I continued to look around the shelter area. The ground appeared to be disturbed and had the appearance that something had been laying down in it. I looked toward the rear of the shelter and lying in the back on a patch of moss was hawk skull. The skull was completely bleached out and preserved with no flesh remaining. It was definitely not there the week before. The skull was in remarkably good condition. The skull was not just randomly dropped here but placed carefully.

I did not know what to do with the skull. I had only been doing "Bigfoot research" for a couple months. I thought maybe this is a place where Bigfoot in this area to hide things so I placed it back where I found it on the moss. I find it interesting that the delicate skull was not broken and was laid on the soft moss. I later checked with some fellow researchers who suggested when I go back I take the skull. That in their opinion a bird skull, especially a bird of prey was meant as a gift and I should take it. So I get the skull on my next visit.

Above is a side view of the skull, notice how delicate it is yet it is not broken. I place it back where I found it, not knowing it was a "gift".

I looked around the area for anymore sign and left the rocks just like I had found them. I wondered what the skull was all about, and what did it mean. I did some research, and to Native American Tribes the Hawk symbolizes a "messenger".

After I finished up around the stick shelter I moved down the horse trail to the area where I had the "Black Nose" Bigfoot sighting to look around again. Each time I visit this area I look for more evidence of the sighting. A fresh perspective helped me think of new places to look for sign of their movement or what area in the brush they may have disturbed. I worked my way around the right checking the upper branches of the cedar tree that the Bigfoot was hiding in. I was able to locate a couple more broken limbs.

The broken limb is interesting because it is approximately where the head and shoulders were located. It's a large limb and it would take considerable force to break it yet there are not large limbs on the ground around the tree that could have broken it. It also protrudes into the smaller fine limbs of the cedar tree. I think the Bigfoot broke this limb as he pushed up into the tree. This is also why I think his shoulders were sideways, and he was turning his head toward me, it was the only way he could fit into the space without making an inordinate amount of noise.

I walked the trails for another 2 hours looking for sign or footprints. I continue to find plentiful food sources. I find nothing else and make my way out of the area without further incidence.

Chapter 6 - The Woods Are Too Quiet

Date: August 22nd, 2009
Location: Research Area A

I go into the woods full of anticipation this day. Will there be a new gift left for me? Will I see the Bigfoot again today? These were the questions running through my mind, but my attention was soon brought back to the woods, it was quiet, too quiet. No birds, frogs, or other animals were stirring when I first entered the woods. Something was there and it had everything on high alert. I move down the trail slowly and hear the faint sounds of footsteps approximately one hundred yards to my left rear. I only hear them when I start moving. When I hear a twig on the ground break and stop the footsteps stop. I think I am being followed and watched. This goes on for about fifteen minutes then a group of crows begin sounding off at something near and I can no longer hear the footsteps.

My first stop of the day is the hide or blind by the water's edge. It has been three weeks since I first found this "blind". I notice when I come upon it that several limbs in the front have been broken and pushed outward. It looked to me like something was lying in the shelter then exited quickly out the front breaking off limbs as it went. There were no large limbs that had fallen to cause all the damage. Several limbs were broken and the broken sections were still attached and pushed upward not downward. I also found two limbs that were at least two inches in diameter.

The Dark Rock Returns

I leave the day shelter and make my way to the "stick shelter" to see if the rocks have been moved or taken, hoping to receive another gift. When I arrive I discover the rocks have been moved. I also find that the "dark" rock I originally left has been returned and the lighter colored rock has been taken back! Did I upset them by not taking the hawk skull?? I was ignorant of what was going on when the skull was left for me. I hope my ignorance has not cost me a chance to continue the exchange of gifts. Below is how I found the rocks. Below is a side by side of the rocks I left on August 16th and how I found them on August 22nd.

Rocks 08/16/2009

Rocks 08/22/2009

In the previous comparison you can see that the light colored rock has been taken and the dark colored rock left in its place. The Bigfoot returned only the one rock, why? What does this mean if it means anything?

The light rock was taken and the dark rock that was returned. Where had this rock been for the last three weeks? Did the Bigfoot carry it around or stash it somewhere?

08/01/2009

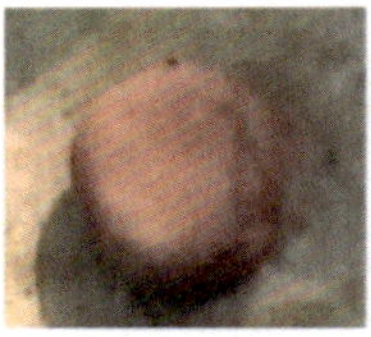

08/16/2009

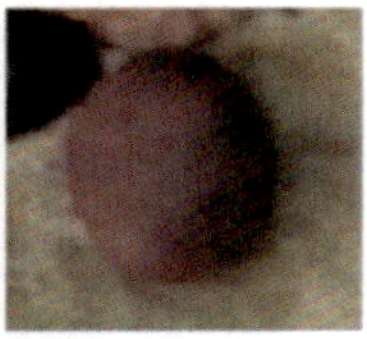

08/22/2009

Above is the time line of the "rock exchange". You can see that the rock I left on August 1st was returned to the shelter on August 22nd. I made sure this time I took the hawk skull in case I upset the Bigfoot by not taking their gift. I also decided to try and leave polished rocks and see if the Bigfoot would accept and take these gifts.

I move down the trail and investigate a cross trail that overlooks the area where I had my August 1st, 2009 Bigfoot encounter. I find a large poison oak vine has been pulled off the side of a Poplar Tree and "placed" into a circle. The vine appears to have been pulled off the tree top down starting at the fourteen foot level. The vine was also looped into a circle and as if to form a blind to hide in. It would be in the perfect location to observe anyone moving across a game trail thirty feet away and the area where I had just been investigating.

Who would rip a poison oak vine off a tree and then form a large circle with it? Touching it and getting the oil on your skin would cause very painful itching sores. Did a Bigfoot rip the vine down quickly in order to form a blind? Was the blind made to observe me or to ambush a deer?

I wrap up this outing and head back to the house. The vine was very interesting but is only the tip of the iceberg as I will find out in the weeks and months to come.

Chapter 7 - Surrounded by Bigfoot

Date: August 29th, 2009
Location: Research Area A

It is the day after my birthday when I hit the trail at the research area. It is early and the woods are quiet. Almost immediately I sense that I am being followed and watched. I move slowly down the horse trail that leads to the stick structure. I hear faint footsteps in the woods to my left. The Bigfoot are just out of sight, but I can hear them. I continue toward my destination cautiously with my "shadow" in tow. I make the right turn at the trail intersection with the faint footsteps still following approximately one hundred feet behind me. I come to the stick structure" and immediately I notice the rocks had been moved. They were in the in a pattern of one rock alone and three together.

I placed the rocks in a straight row on August 22nd, 2009 in the picture on the left and the rocks are moved into the 3-1 pattern on my return on August 29th, 2009.

The previous visit I had set a "stick trap" by laying out some sticks on the ground to see it would get moved or stepped on. When I arrived at the stick structure it was obvious the ground had been disturbed and the stick trap "sprung".

I begin to look around the area for other trace evidence when I hear a noise in the brush behind me. I quickly turn around and point my camera toward the direction of the sound. After about three or four minutes I hear another sound of something moving in the brush behind, so I turn and look in that direction. I realize I am being played. I decide to go ahead and continue my investigation of the site and put out some peanuts that I had brought. Now that I am ignoring them, the noises become almost constant. I am literally surrounded by Bigfoot. When I looked in one direction I pointed my camera in the opposite direction. I did this in case I was being distracted so they could take a closer look at me. My hunch was correct and I caught one popping its head up only ten feet away! It had dark brown/black hair and black skin. It apparently had crawled through the underbrush then raised its head up for a few seconds. Though I was not looking in the direction of the Bigfoot I was able to capture a few frames of it. The interesting thing is that the Bigfoot pulls vegetation in front of its eyes. This is a tactic I have seen them use before. They know that recognition happens when there is eye contact.

In the above still capture the Bigfoot has raised its head up through the vegetation

While working on the enhancement of the video I noticed a couple frames where the fingers of the Bigfoot were visible. It appears that it has pulled some vegetation out of the way so it could look at me. In the photograph below the end of the fingers are visible.

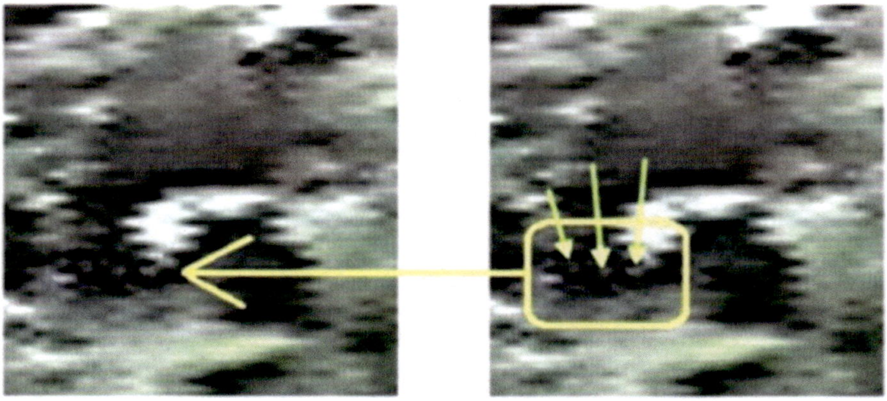

In the next set of still captures the Bigfoot turns its head slightly to the left and a small ear becomes visible.

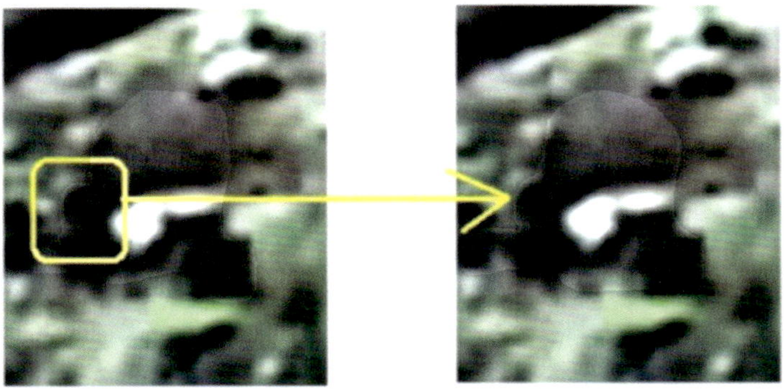

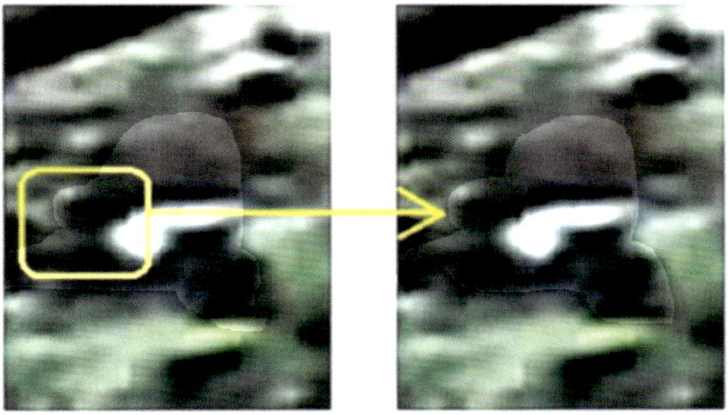

In the below photograph the nose of the Bigfoot is visible

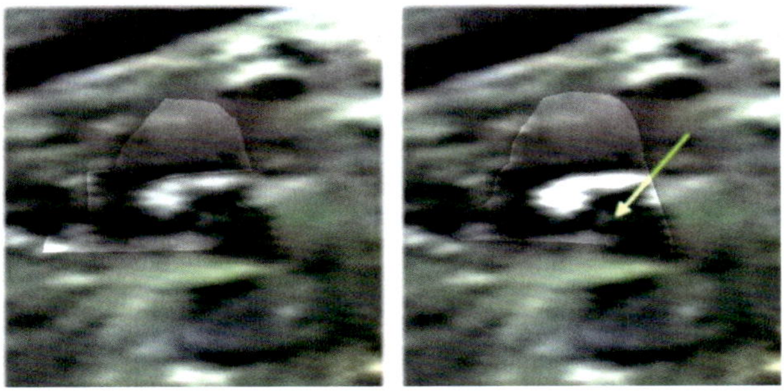

I was distracted by other sounds and only captured thirty frames of the Bigfoot raising its head up. A few minutes later I filmed the same area and the Bigfoot was missing. On the next page the photograph on the left shows the Bigfoot looking at me, and the photograph on the right is the same area approximately five minutes later. I turned my back, the Bigfoot slipped away, and I did not hear the Bigfoot make any noise as it crawled away.

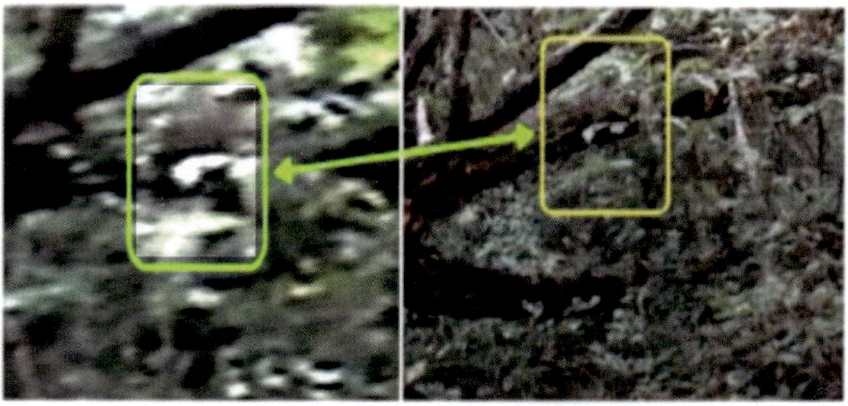

The drawing below is my interpretation of the Bigfoot

I then get the roasted peanuts out to leave for the Bigfoot. I am trying to gain their trust and with my limited experience at the time I thought leaving roasted peanuts would show my intentions as being friendly. I get up and move around the back of the tree where my back pack is laying. As I move around I aim my camera to my left toward a brush pile. I was not looking in the direction I was filming but the camera captured a few seconds of a Bigfoot sticking it's head through the brush to take a look at me while I had my back turned. I call this Bigfoot "Gray Face" because it has a distinctly gray face. I think the digital camera algorithm also helped turn the entire head gray including the hair. This is a drawback of digital cameras and can sometimes distort the true color of an object.

Below is a still capture of the Bigfoot I have named "Gray Face"

When I first saw the above image while reviewing the video I was shaken. I had yet to find the little one that had crept up so close, but this was a full facial view and I had nightmares for a couple weeks after finding this on my video.

As I mentioned before I have spent many hours working on the enhancement of the seventy two frames I captured. I am amazed how deep set the eyes are. The eyes cannot be viewed looking directly at the Bigfoot from the front. I did some research and discovered this trait has been widely reported by eye witnesses and major research groups.
(BFRO – http://www.bfro.net/gdb/show_FAQ.asp?id=585).

I used photograph enhancement software to run the eye area through many filters trying to bring out any detail in the eyes. Below is an interesting close up of the eyes.

The eye close up, notice the detail of the pupil and the iris

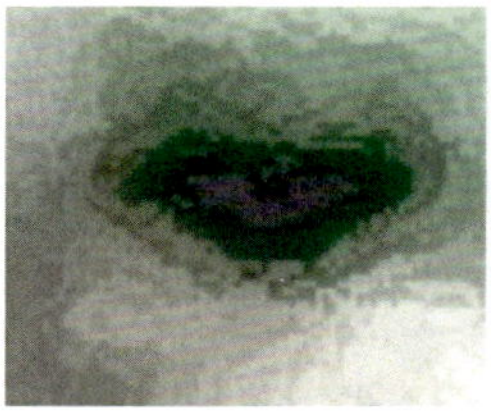

I also noticed when studying the clip in slow motion that the mouth of the Bigfoot was moving. It looked like it was possibly chewing on something.

"Gray Face" is what I call this Bigfoot. I think it is female; this is just based on my gut feeling not any tangible evidence. When I followed up the next week to make sure this was not an anomaly or artifact created by shadows and limbs the area was empty. There was nothing in this area.

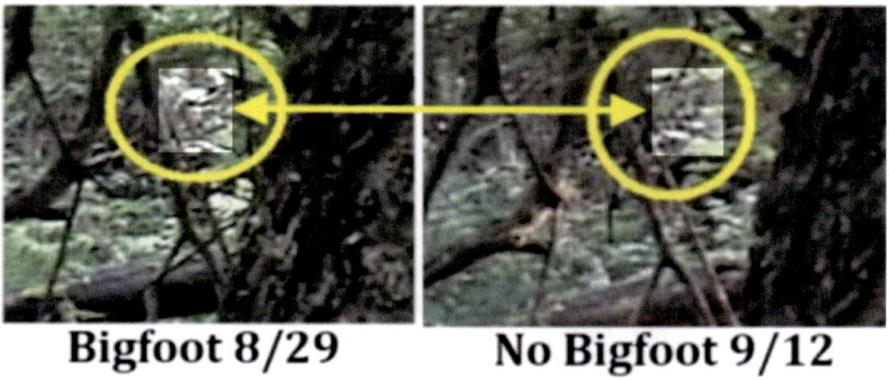

The below sketch is based on the still captures of the "Gray Face" Bigfoot

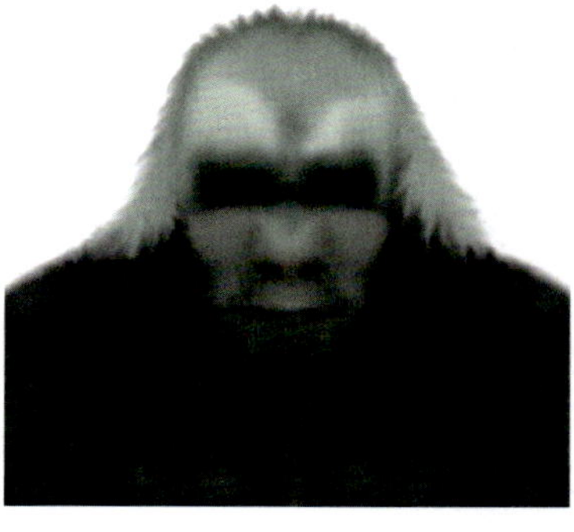

The Bigfoot are playing me like a fiddle. I was ignorant at the time of what they were doing to me. It was a classic hide and seek. I was surrounded by at least three maybe four. They would take turns distracting me by making noise. When I would look one way then behind me one would stick out its head and take a peek at me then quickly hide again. I even had one belly crawl within ten feet of me to rise up its head and take a look, then quickly slip away. They had me nervous and paranoid. I was constantly turning around in circles. The rest of the day I was followed loudly. They wanted me to know I was being followed. They stayed about one hundred feet back and quartering to my left or right.

The big shock for me came when I was reviewing the video and finding I was being watched. This was a good lesson for me on how the Bigfoot behaves.

The Nightmares Begin and End

Up to this point in my research it has been a "romantic quest" to find the elusive Bigfoot. I had been lucky, in the right place at the right time. I did not realize what I was actually dealing with. After the entire experience of this day the gravity of what I had experienced hit me. I had been within a few feet of a person that was bigger, stronger, and unpredictable. The Bigfoot could have attacked me and ripped me to pieces at any time. Then I considered the "Gray Face" Bigfoot taking a look at me and the possibility of it being female. I wondered if the smaller Bigfoot that sneaked in so close was its offspring. What if I had noticed it, and moved toward it, would she have attacked?

All these questions and more kept running through my mind. This was the event that changed the way I looked at the Bigfoot. Gone was the child like fascination and it was replaced with a new found respect. The Bigfoot could be within feet of me at any time and can kill me with ease. I would never look at the Bigfoot in the same way again.

I also began to have nightmares about Gray Face attacking me. This lasted a couple weeks. It took a while to process what was happening and to come to terms with it. This was not an adult game of hide and seek I was playing. These are real people, extremely intelligent, masters of stealth and survival. They have the advantage, and I was the novice, a mere child amongst giants.

I was scared to go into the woods for the first time in my life. I had always felt at home in the wild. I loved the woods and just being out in the woods. Knowing that the Bigfoot exists changed all that. In a sense my innocence was lost. I could never go into the woods and know I was alone. Hunting, hiking, and camping would never be the same. Now when the hair stood up on the back of my neck I would know why. When unexplainable things would happen while I was out on a hike, I would know why.

It took time, but I finally came to terms with the fear. I turned to my faith in God, and the knowledge that he is in control of everything. I do not believe in luck or coincidence, and if God was in control I have nothing to fear. This calmed my spirit and gave me the inner peace about going back out into the woods again. I will not lie, I do have a healthy respect for the Bigfoot, but I no longer fear them and can enjoy the woods again.

Chapter 8 - First Trail Cam – Bigfoot watches me as I follow-up from previous weeks sightings

Date: September 5th, 2009
Location: Research Area A

I purchased my first trail camera to try and get pictures of a Bigfoot. A trail camera can stay in the woods twenty four hours, seven days a week. They have motion sensors that trigger the camera to take pictures. The model I purchased has the ability to take both still shots and video. It also has infrared emitters for taking pictures and video at night. A trail camera will give me the ability to have a presence in the woods around the clock.

Trail cameras can also have some draw backs. In researching the use of trail cameras for Bigfoot research I discovered that a major issue with the cameras is the electronic emissions. The Bigfoot appears to have the ability to hear the cameras. The cameras also smell because of the plastic used in their construction.

My solution to these issues was to encase the trail cam in a custom metal box I built from an ordinary cash box. I also used egg crate foam to insulate the inside to help further reduce the electronic emissions. The steel case is also lockable, and will help prevent theft or destruction due to bears or vandals.

I setup the game camera and then travel to the stick shelter" to check on the gift exchange project, and do some detail follow-up on the two Bigfoot I had captured the week prior on video. I wanted to take video from the same angles to make sure the faces were not caused by shadows or a combination of sticks and leaves. I also wanted to look for trace evidence like hair, footprints, disturbed vegetation and broken limbs.

When I arrived at the stick structure the peanuts I had left the week before in a white handkerchief were on the ground. There were none missing so I chalked it up to squirrels. Next I went about documenting the areas where the Bigfoot had been the week before.

I continued to work around the area taking close up shots of where the Bigfoot had crept up so close to me. I was careful to look for hair and disturbed vegetation.

While I was moving around taking follow up videos I was unaware that I was being watched from the brush. When I returned home and reviewed my video I found a Bigfoot had been watching me.

The below photograph shows the Bigfoot looking and me with close up inserts

Composite Photo of the Bigfoot

I used photo editing software to make a composite photograph of the Bigfoot's complete face. Since I have at least half the face, I will flip it creating both sides of the face. Then I join them together to give me an idea what the Bigfoot's full face might look like.

In the below composite the nose, cheeks, eyes, pronounced forehead, deep set eyes, and ears!

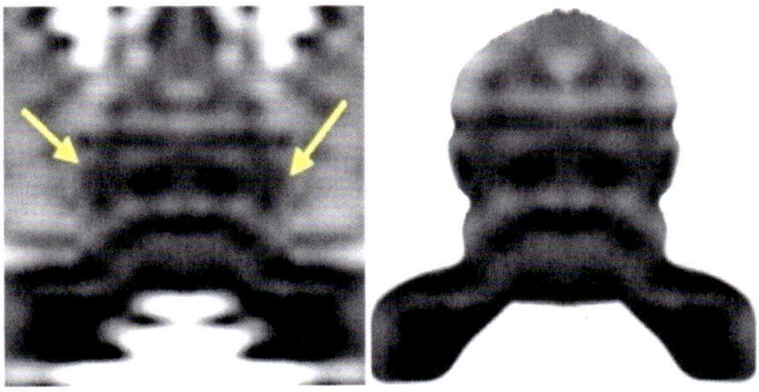

I was never aware of this Bigfoot's presence. I moved around the site for a few minutes then took some more video of the area and the Bigfoot was gone. During review of the video I discovered that the Bigfoot were not present in the location just five minutes before being captured on video.

The below photograph shows the Bigfoot not present five minutes before it was captured on video

The Bigfoot is watching me in the below still capture as I document the location

The below still capture shows the Bigfoot missing five minutes after I captured it on video

Rock thrown at me as I head down the trail

After finishing up at the stick structure I went down the trail. I began hearing something following me in the heavy brush to my left. I soon had the very strong feeling I was being watched again. I may have even been experiencing infrasound. The feeling was so strong that I stopped and turned toward the thick brush on my left, and began looking to see if I could find the Bigfoot that was following me. Then at that moment, a rock came flying out of the woods, and hit the trail just a few feet behind me. I had my video camera running and captured the incident on video. I immediately scanned the woods where the rock came from looking for where the Bigfoot was hiding. It was obvious he was trying to get my attention.

After I surveyed the area for any sign of a Bigfoot watching me I head toward the point where I had first had the experience while fishing. When I started walking out an old and overgrown logging road I began to find sign that something very large had recently walked this trail. Limbs on the ground up to 2 inches in circumference had been stepped on and broken. I walk down the point on an old overgrown logging road and find sign that something large had walked this road recently. Some of the sticks broken were over two inches in diameter and I had trouble snapping them even when using my leg for leverage. I had the gut feeling I was being watched while I was exploring the "point". I would hear the sound of footsteps shadowing me from time to time, but could never see anything. I think the Bigfoot can be almost silent when they want to be. I also think that they will make noise to let you know they are following you. I also think this is what was happening to me this day. This would end my eventful outing for this day.

Chapter 9 - A Mushroom, a Watcher, and Footprints in the Mud Holes

Date: September 12th, 2009
Location: Research Area A

I am anxious to get back to the woods. I have placed my trail camera on a trail that leads from the area I think the Bigfoot are bedding during the day to the agriculture fields of corn and soy beans they are feeding in during the night.

I worked my way into the woods and then took the side trail where my camera is located. I was excited with the prospect that the trail camera may have captured a Bigfoot. I was very naive and in my mind I could just see a clear sixty second video of a Bigfoot walking by my trail camera. I would soon be educated on the abilities of the Bigfoot to hear trail cameras and see the infrared light that illuminates from the cameras infrared emitters.
I begin walking up the trail and immediately start finding signs that something large has walked the trail since I had set my cameras a week ago. I felt a huge adrenalin rush as I examined the possible footfalls.

After documenting the footfalls I continue to move deeper in toward my trail camera. I just "know" with all of the footprints on the trail I have captured a Bigfoot on my trail camera. When I approach the trail camera I notice something on the ground that was not present when I put the trail camera on the tree. There is an "X" on the ground in front of the tree where I have the trail camera mounted.

I found this odd, but continued about the business of getting my laptop computer out so I could view the videos captured by the trail camera. I just knew now there "had" to be something on the trail camera videos! I thought to myself; who would I call first? What would I do? The computer was now booted and I connected the USB cable. I opened the SD card folder and to my shock and disappointment there was not one video clip captured. I had just learned lesson one in trying to capture a Bigfoot on a trail camera, it is almost impossible! I think they can hear the camera's "Passive Infrared" (PIR) motion sensor. They heard the camera and marked the trail camera with an "X" on the ground.

While I was checking the trail camera I placed my video camera on an old fence post. I pointed the video camera so it would record what I was doing and watch my back. When I was reviewing the video I noticed something in the back ground that appeared to have deep set dark eyes and blended in to the foliage. I noticed that the subject was slowly moving left to right. It also tilted its head to the right then back to the left. The subject remained in the same location for the entire eight minute video clip.

In the below still capture the yellow square marks the area where a Bigfoot is watching me download the SD card from my trail camera.

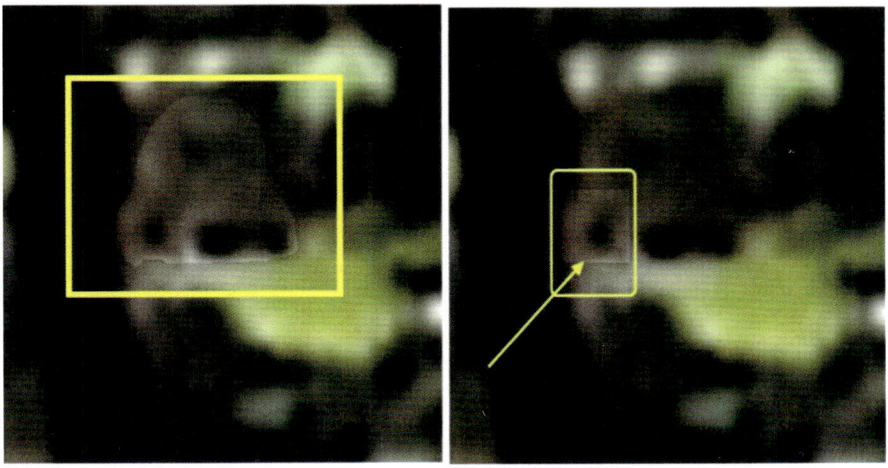

Above is a close up of the subject at the 2:45 time stamp. Note the apparent ear marked by the yellow box.

The following is a comparison of the position of the face from different time stamps during the video.

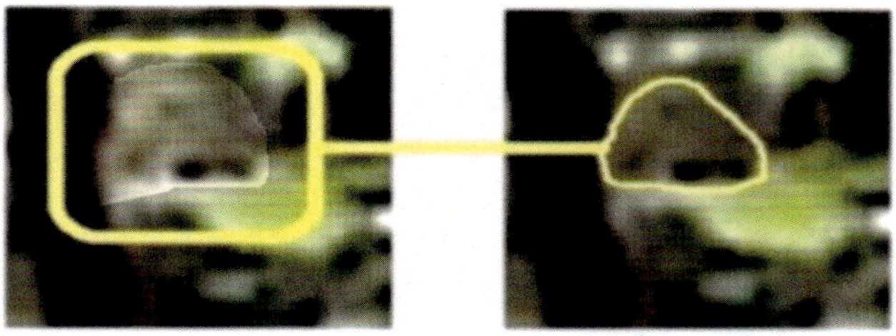

Time Stamp 2:25

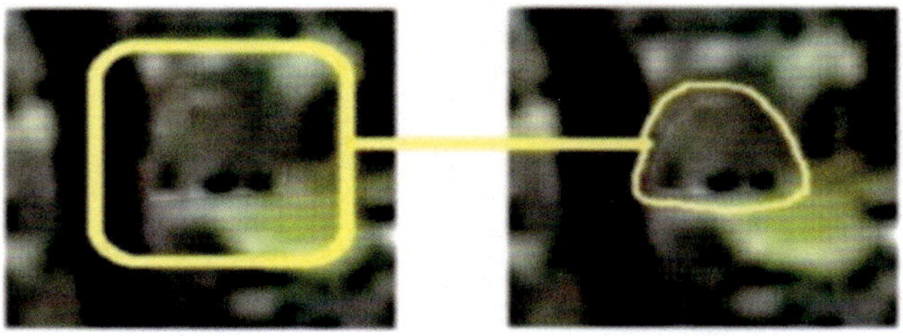

Time Stamp 5:30

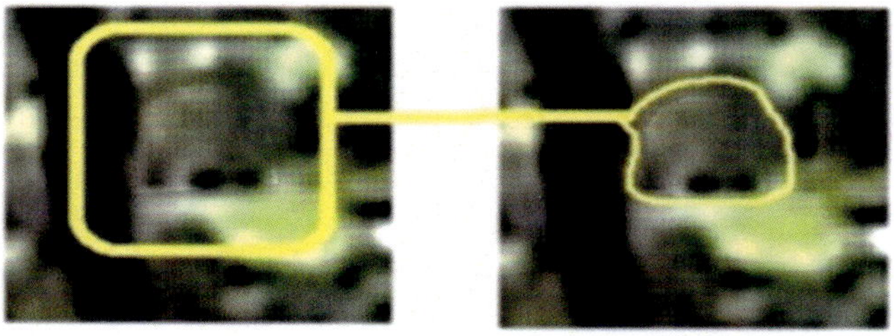

Time Stamp 8:45

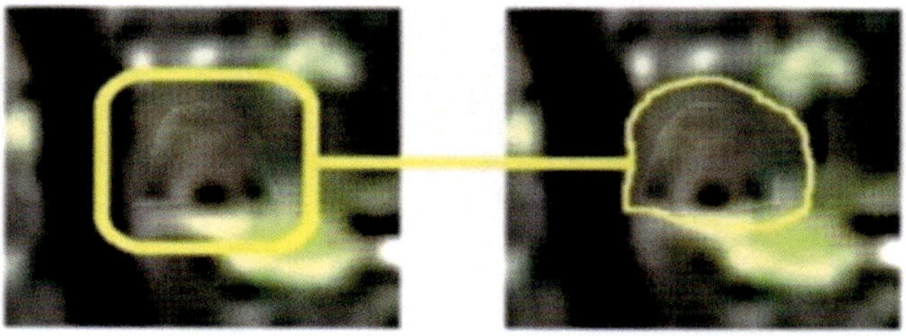

Time Stamp 11:09

I felt like I was being watched but could not tell from what direction. Like many times before I seem to always capture them when I least expect it and when the camera is pointing where I am not looking. The characteristics of these photographs demonstrate the deep set eyes. Many witnesses have reported that the eyes are deep set and almost impossible to see straight on. The coloration of the head and face blend into the background. No wonder the Bigfoot is so difficult to find.

After finding no footage of a Bigfoot recorded on the SD card and with the Bigfoot apparently marking this camera location with a "X" on the ground; I decided to move the camera to another location. There is a game trail that runs behind the area where I had the sighting on August 1st, 2009. I decided this would be a good location to place the trail camera. I started loading the camera into my back pack when I noticed the "Battery Low" light on. I had forgotten the batteries, so I had to go back out to my truck to get them. On the way back to the truck is when one of many interesting coincidences happens. Approximately half way back I noticed a very delicate mushroom "placed" in the middle of the trail. I know the mushroom was not there on my way in earlier in the morning. It was completely intact, and looked like it had been plucked, and then laid in the trail for me to find.

Below is the mushroom laying in the trail

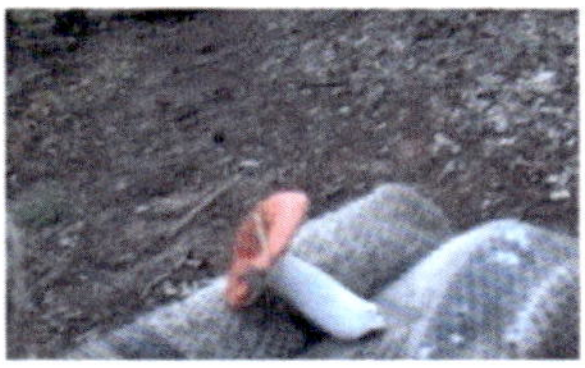

I pick the mushroom up to take a better look at it. I am puzzled. I ask myself "where did it come from?" I am looking around for where it could have come from and in this area of the trail there are no mushrooms.

This mushroom is so delicate that it fell apart by the time I arrived at my truck just a few minutes later. I had placed it in a plastic zip bag and put it in my shirt pocket to "protect" it but that did not work. How did something so delicate blow or fall into the trail? I think it was placed in the trail for me to find? The question is why? Is this a gift, I'm sure this is a delicacy for the Bigfoot, a small tasty treat. I would like to think that is what it was a gift. I know by this point in my research in this area that the Bigfoot clan has "scouts". They locate and follow potential dangers. They also warn the other members of the clan. The scout for this clan, whose picture is probably in the previous montage, began to let me know that it was following me. When I enter the woods almost immediately I am followed. The scout always stays about 100 feet behind me quartering to the left or right. The scout began making more noise as it followed me the more often I would visit the area. This is conjecture, but I think the noise making is to let me know they are

watching me. Then I begin getting "gifts" like the mushroom and I hoped in their way the Bigfoot were making "friends" with me.
The term "friend" I use loosely because the Bigfoot are so reclusive that this means little glances, gifts, and making noise so you know they are there.

I retrieve the batteries from my truck and reset the camera. Then I head to the area where I am going to set the trail camera. Going down the trail I hear the tell tell noises of bipedal walking near me, shadowing me. This does not concern me but gives me an odd sense of acceptance that I am followed and not run off.

When I encounter the usual mud holes on the trail I notice what looks like something had walked through the middle of the mud hole. Upon closer inspection I recognized what look like large toes that have been sliding in the mud. There was also evidence of the toes spraying as the Bigfoot was struggling to make its way through the mud hole. I find two footprints one from a left foot and one from a right foot. It appeared the Bigfoot was walking in the previously made depressions in the mud hole to hide its footprints. The Bigfoot will take very ingenious measures in order to hide any sign of their movement. Many times I find deep impressions or broken vegetation on the edge of mud holes or trails that are soft enough to leave footprints. The size of this mud hole and the very thick vegetation on both sides made skirting the edge impossible. The Bigfoot then did what made sense, walk in the center of the mud hole in the prints and holes previously made by the horses. I was fortunate to come along not long after the prints were made. It would not take long for horse traffic and time to remove all signs of the Bigfoot's passing. This goes to demonstrate that the Bigfoot are very intelligent and can easily solve problems to stay stealthy and hidden.

This is footprint number one that was made with the right foot. In the photograph below you can see the big toe as it separates and the four other toes as they slide back down into the mud.

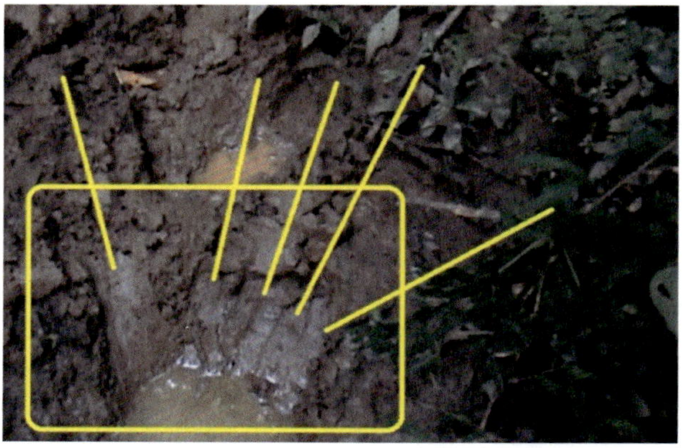

This is footprint number two made by the left foot. The photograph below shows the five toes and the mud that squeezed between them as the print was made.

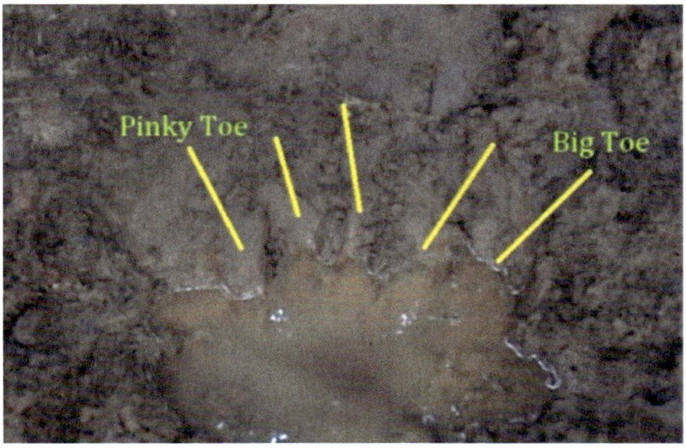

I navigate around this mud hole after documenting the prints and head to the area where I plan to place my trail camera near the August 1st sighting area. I placed the trail camera to look down the trail to give maximum exposure time if something walks down the trail.

The below still photos demonstrate the visual camouflage of the camera. In the photo on the left the camera is not marked and the photo on the right the camera is marked. I thought it was difficult to see them standing just a few yards away from them. I was very excited but again in time I learn that visual camouflage does not fool the Bigfoot.

Once I complete the trail camera placement I make my way toward the stick structure site to see if the rocks have been moved around or I can find any sign of activity. When I get close to the site I have to work my way around a large mud hole. I have found footprints in this mud hole before and again I find a large footprint in the mud hole. This footprint has a very interesting feature. It appears that the big toe stayed extended while the other toes curled downward to grip. You can also see where the toes pushed up mud as the foot was lifted out of the mud. I did not measure the print but I would approximate its size between eighteen and twenty inches.

I mentioned earlier that you could see the mud that had been pushed up by the toes as the foot was picked up. In the below still capture I point this out. Also note that the big toe looks to have stayed extended while the other toes curled to get a better grip. Is this a genetic trait? Has this Bigfoot been wounded and can no longer move or use the big toe? These are interesting questions.

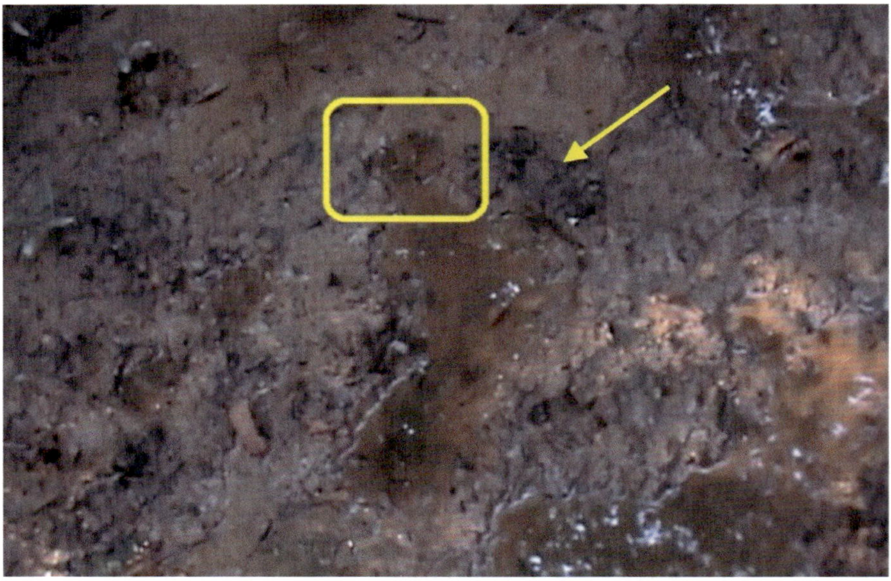

After I document the footprint in the mud hole I work my way to the Stick Structure location where I find that something has disturbed the shells I left the week before. The 3-1/4-1 Pattern continues to be repeated at the Stick Shelter location.

Below is the comparison of how the shells were left on September 5th, 2009 and how they were found on September 12th, 2009.

Shells on 09/05/2009

Shells 09/12/2009

Note that the shells are placed in the pattern of one item being placed separate with the others together. In this example the pattern is 3-1 with one shell moved off the rock. I considered that this could be a coincidence, but as you will see throughout this journal this pattern persist. After documenting the shells I left the area for the day and headed home.

Chapter 10 - A Soggy Day, Candy Bars, and Garlic

Date: September 20th, 2009
Location: Research Area A

I woke this day and looked outside and it was pouring rain. This was a day better for sleeping in rather than going out and doing Bigfoot research. I almost stayed home but I just had a gut feeling I should go. I had been doing research and found through other experienced researchers that Bigfoot seemed to have a sweet tooth. So I purchased a couple candy bars to leave for the Bigfoot to try and build some trust.

I also did some research concerning a Bigfoot habitation case in Madisonville, Tennessee. Janice Carter Coy claims that for fifty years she and her family had a Bigfoot family living around and on their farm, and her family had been interacting with them. Janice Carter Coy wrote a book about her experiences that was co-authored with Mary A. Green, called 50 Years of Bigfoot: Tennessee Chronicles of Co-Existence. This alleged habitation was only ten miles from my research area. I have read reviews on the book and there are very diverse opinions on the validity of this case. There is a definite split in the Bigfoot community on this case. Many think that it was a complete fabrication while others think it was a true and legitimate habitation. I had not read the book yet, and would not read it until March of 2011. Since it was so close to my research area it is reasonable to conclude that if any of the claims made by the book are legitimate then these Bigfoot I have been interacting with could be related to this group. I had recently watched a TV special on Bigfoot where Janice Carter Coy told that the Bigfoot asks her for Garlic. So, I thought I would try leaving some garlic along with the candy bars to see what would happen.

I packed my gear, put the video camera inside a plastic bag and off to the research area I went. The rain was constant and once I arrived and started down the trail it was evident it was not going to let up.

The trail is completely underwater. I had made up my mind that I was going to leave my gifts and make a plea to the Bigfoot. So I continued on in the thick mud and heavy rain.

When I arrived at the Stick Shelter I noticed that the shells on the flat rock had been moved some. The movement is subtle but again the pattern is one by itself and the others grouped together. This is the fifth time this pattern has been made. I think a reasonable interpretation would be I am not part of the group. The Bigfoot are together and I am separate. After documenting the shells I rearranged them and moved on to setting out the candy bars I had brought.

I then turned my video camera off, and spoke to the Bigfoot that I hoped were watching me like they have in the past. I had never talked to the Bigfoot before in this way. I did not know if they could even understand me, but I wanted them to know I was not a threat, and wanted to interact with them. Many people want to befriend them, and have the romantic notion that the Bigfoot is the "guardian" of the woods and nature. I do not think this at all. They are a living, breathing people who are surviving in their environment. They have the three basic needs we all have food, water, shelter. I was hopeful that by providing something they needed that I could begin to build a reason for them to trust me and interact with me.

I then moved back out to the main trail to investigate the path that leads to the small island. I wanted to see if anything had happen to the stick I put across the trail where it comes up out of the water.

The week before on September 12th, 2009 I had placed a stick across the trail where it entered the water going over to the small island. I placed it just off the ground about twelve inches. When I return on September 20th, 2009 the stick has been knocked to the ground. It was not moved toward the island but away from the island. This lead me to the conclusion that whatever had hit it was coming across the shallow water from the island and moving toward the main land. This also meant that whatever hit or kicked the stick had to be on the island when I placed the stick across the trail. I think it was a Bigfoot, but that is my theory, a gut feeling if you will, but one that makes sense given the history of this area.

The stick across the trail on 09/12/2009

The stick knocked down and away from the island 09/20/2009

Above on the left is the stick I left across the trail on September 12th, 2009 and the picture on the right shows the stick on the ground and about two feet inland on September 20th, 2009.

I think that the Bigfoot kicked the stick as it stepped up onto dry land after crossing the shallow water that divides the island from the main land. After documenting this I made my way back toward the truck and went home. I was hopeful that my "gifts" would be accepted.

Date: September 26th, 2009
Location: Research Area A

I had been grappling with the decision to place trail cameras at the stick structure location for a couple weeks. I had been posting my findings on YouTube. Many people were questioning why I had not placed a trail camera in the location in order to catch on video what was moving the rocks around and leaving me gifts. I also had other researchers telling me that placing a trail camera in this location would stop any activity. The Bigfoot had an uncanny way of sensing that the trail cameras and avoiding the area. One author even wrote if you are having trouble with Bigfoot bothering you at your home then put up a trail camera. This will cause them to leave.

I have a couple of theories why trail cameras have this effect on the Bigfoot. The most likely reason in my opinion is that they can hear the motion sensors on the camera. I also think that many times they are watching when you place them and they can also see in the infrared spectrum. If a trail camera is tripped the infrared would look like a bright light to the Bigfoot.

It was a tough decision but I thought at the time it was worth the risk if I could get a Bigfoot on video. So I purchased my second 3.0 megapixel trail camera. It was the best I could afford at the time, but a good choice for the money. The cameras are easy to configure and reliable. Battery life was not as long as I would like, only about two weeks, but it was acceptable given how close I lived to the research area.

I headed out very early and arrived before day light at the research area. When I rounded the corner to park I noticed a small beagle mix puppy in the middle of the road. He was a pitiful looking thing, dirty, hungry, and flea bitten. He had obviously been put out by his owner and left for dead. When I parked he cautiously came over to me and gave me the classic puppy dog look. I tried to ignore him and go about getting ready to hit the trail, but my conscience got the best of me. I picked him up and put him in the back of my truck bed. I told myself if he does not jump out I will take him home. Of course he was so little I knew there was no way he could jump out.

I made my way down the trail toward the stick shelter area as the sun began to rise. It was still not light enough to film so I had to use the infrared feature on my Sony HandyCam video camera. This feature uses an infrared light and lens filter to capture video in the dark or low light conditions. I am anxious to see if the candy bars and the garlic were taken and any gifts were left for me. The candy bars are missing and the garlic has been scattered on the ground.

I gathered the garlic back up and placed some more candy bars I brought with me back in the tree.

Since I would be placing the trail camera watching this area I was hopeful that the candy bars would help lure a Bigfoot in front of the trail camera. I also checked the small rocks and they have not been disturbed. I placed larger rocks in their place because several critics had claimed that the smaller rocks and shells could be moved around by small rodents or inclement weather. I decided to place heavier rocks out to eliminate this possibility.

After setting the rocks I set up the trail camera to monitor this location. I placed the camera back into the brush and approximately three feet off the ground. I aim it into the stick structure hoping to capture anything that tries to take my bait or move the rocks around.

After setting up the trail camera I begin to get the feeling that I was being watched. I slowly panned the camera around me in a full circle. The light was very low so I was still using the camera in infrared mode. I did not see the following at the time but upon review of the video I noticed eye shine that was being caused by the infrared light. I also missed a large Bigfoot standing to my left and a third Bigfoot hiding in the brush closer to me than the one giving off the eye shine. I think that the entire clan was watching me put up my trail camera.

In the above still capture you can see the eye shine made by Bigfoot number one

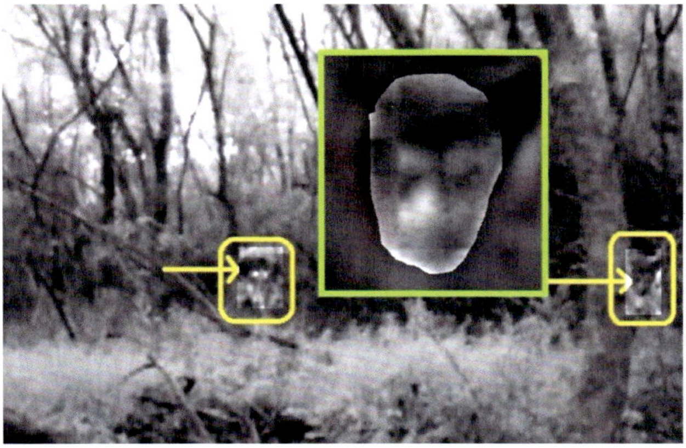

In the above still capture two Bigfoot are watching me

The final Bigfoot that I captured on video that day is in the above still capture

I had the feeling I was being watched and I was constantly looking around. The video captured the Bigfoot I sensed were around me but could not see. It was at the break of dawn and the conditions were low light. I was not looking through the view finder, I was holding the camera at chest level and as I would look in one direction I would point the camera in the opposite direction. I think this explains why the Bigfoot were showing themselves. The Bigfoot do not know what end of the camera contains the lens. They observe the direction I am looking and assume the camera is pointed in the same direction.

There was so much going on so let us begin with the Bigfoot that was reflecting the infrared light with its eyes.

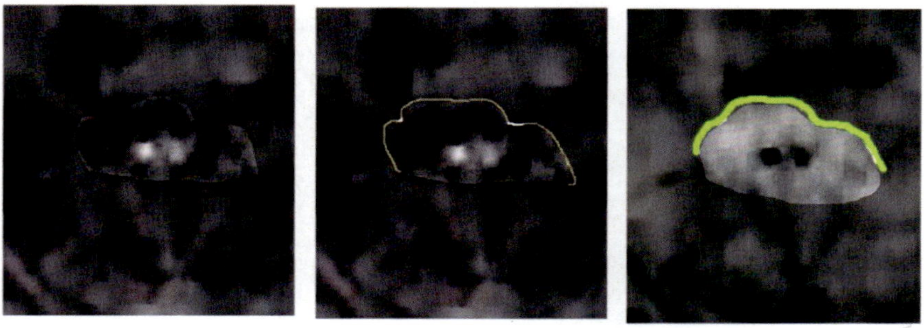

In the above still capture the head, eyes, and shoulders are visible. It appears this Bigfoot is on its hands and knees in order to stay concealed. Notice how wide the shoulders are, it resembles a football player in shoulder pads. The eye detail is visible in the negative still capture on the far right.

The second Bigfoot was very close and was either lying down or squatting. Its face was visible at the same time the other Bigfoot eyes were illuminated.

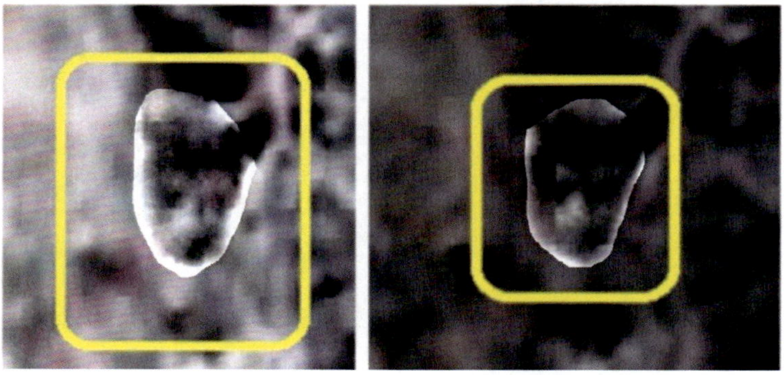

The final Bigfoot, that was watching me this day, was hiding to my left peeking out from behind a tree. I think this is the Bigfoot I have named "Black Nose". He is approximately eight feet tall. In the video you can actually see him move as I turn my head to look in the direction I have the video camera pointed. The Bigfoot have exhibited this behavior before, sticking their heads out to take a look while my head is turned then ducking back into cover when I look in their direction.

I went back to the previous week's video and was able to find a section of video where I was filming the exact same area where the Bigfoot was standing. Above is the comparison to show that the Bigfoot is not a tree, dark brush, or some other anomaly. The Bigfoot observing me place the trail camera would explain many of the events that would follow in the coming weeks.

Below is a close up of the Bigfoot's profile

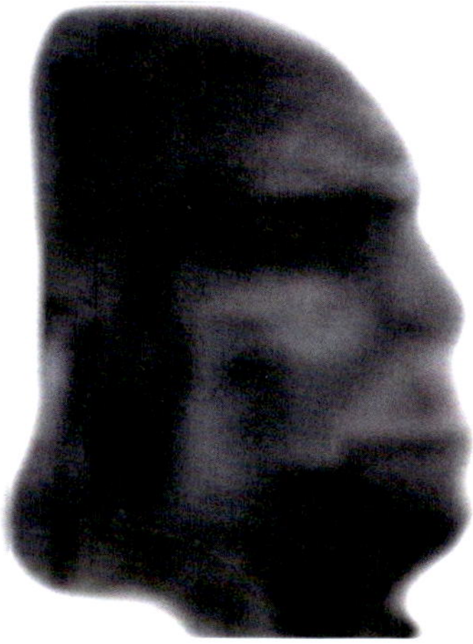

Take note that the Bigfoot's facial profile resembles a Native American in profile. This would be the only events worth noting during this trip to the research area. I complete the rest of my day without incident and return home.

Chapter 12 - Being Watched, Footprints, and Face on the Trail Camera

Date: October 4th, 2009
Location: Research Area A

I was excited because today I was going to check my trail camera to see if I had captured any video of a Bigfoot. It was also bow season for deer so I could get in a little hunting as well. I arrived at the woods early and started making my way down the horse trail. Almost immediately upon entering the woods I began hearing bipedal walking sounds off to my left. I could not see anything but this was becoming a familiar routine with the Bigfoot in this area. I would enter the woods, then the one I call "Scout" would pick me up and start following me. He would "tip his hand" by making some noise to let me know he was there and following me.

I eased down the trail with my escort in tow when a deer was spooked to my left in approximately the same location the Bigfoot was walking as it followed me. Seems not even the Bigfoot can fool everything all the time and this deer had busted him. The deer snorted and blew several times to sound the alarm to other deer in the area. She made a fuss for about two or three minutes then retreated into a thicket for safety.

I continued on down the trail still looking to my left trying to see if I could spot my escort when I got the familiar feeling again. I was being watched by something behind me. I turned around very quickly and got a couple seconds on the video of a Bigfoot watching me from behind. I wondered if this was my escort or another Bigfoot since it was almost fifty yards directly behind me. Of course with the speed and agility the Bigfoot has it would have been an easy task to cover the two hundred yard distance and get behind me on the other side of the trail. In fact that makes sense once the deer had exposed the Bigfoot's location.

In the still capture above I highlight the area where the Bigfoot is located

The face is green because of the algorithm the digital camera uses when recording the image. A digital camera will convert the image into bits of data. It has to decide what colors to record. If you have something tan colored surrounded by green then the camera will store the area as green.

A "Face" captured by the trail camera

After I documented the footprints I spent the rest of the day deer hunting. It was not until I returned home and reviewed my trail camera videos did I realize why my trail camera had been moved. There was only one video on the camera but it captured what I think is a Bigfoot. When the video first begins something is shaking the camera. Then a hairy arm and hand pass in front of the camera causing an infrared flash from the reflection off the arm.

The infrared clip is in black and white. Flesh will reflect the infrared light and look pasty white. The next still capture shows the face. This is one of those situations where you either see it or you do not. When I posted this clip on YouTube the reaction was mixed. Some people saw the face very clearly and could also see it moving slowly. Other people saw nothing and had a very visceral reaction to the video claiming the face was the result of shadows, leaves, and a matrix affect.

Below is a still capture of the "Face"

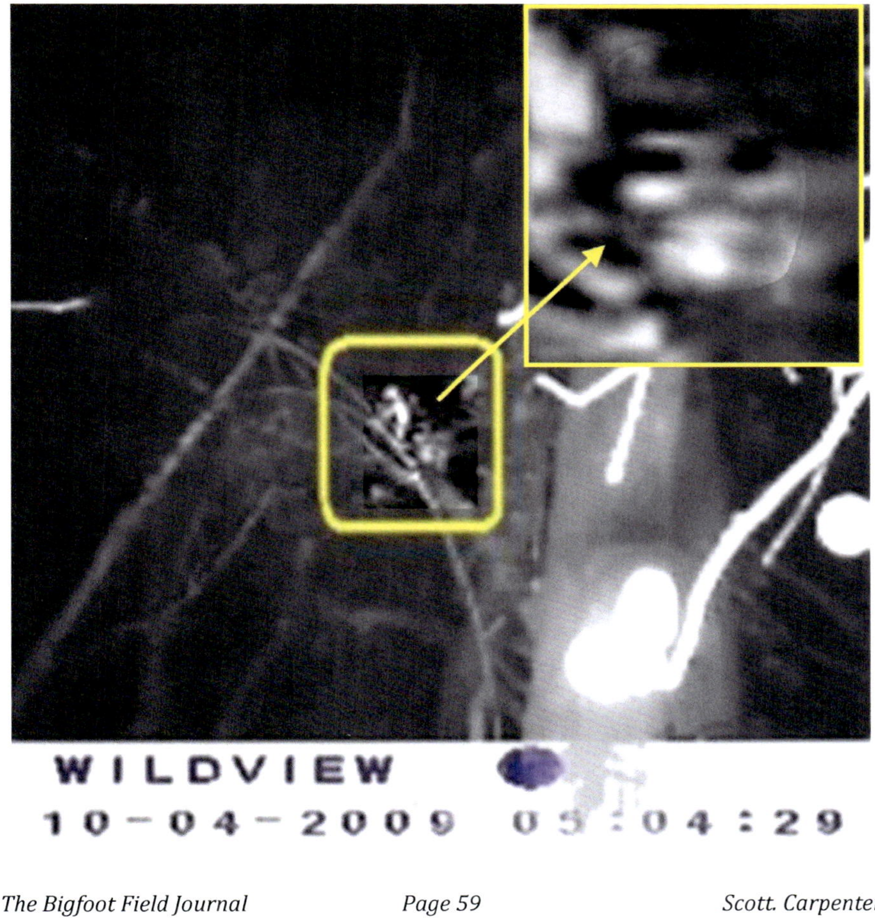

The critics of this video clip claimed that the face was nothing more than a matrix affect combined with shadows. That my brain "makes" the face and there is nothing there. I think this logic is flawed for several reasons.

The critics totally ignored that the camera was shaken and a large hairy hand and arm was waived in front of the trail camera.

The face moves several inches. At the beginning of the video the face is looking to its right, and then slowly moves to the left, then finally back to the front looking straight at the camera. There was no visible wind and the camera was fixed. It was actually strapped and cable locked into position so the camera did not move except when it was shaken in the beginning.

The eyes can be seen in detail, and when the arm passes in front of the trail camera it causes an infrared flash due to the reflection back off the camera. During this flash the iris of the eye shrinks to a pinpoint size. A few seconds later when the infrared light returns to normal output, the iris of the eye adjust and dilates.

In the below set of still capture the movement of the face is demonstrated

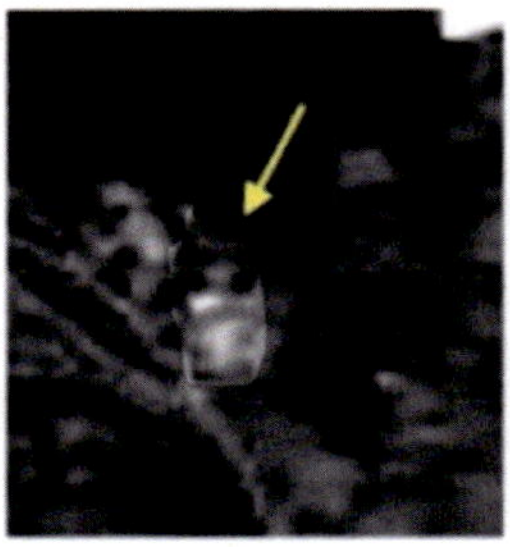

Looking Right
Time Stamp 10 sec

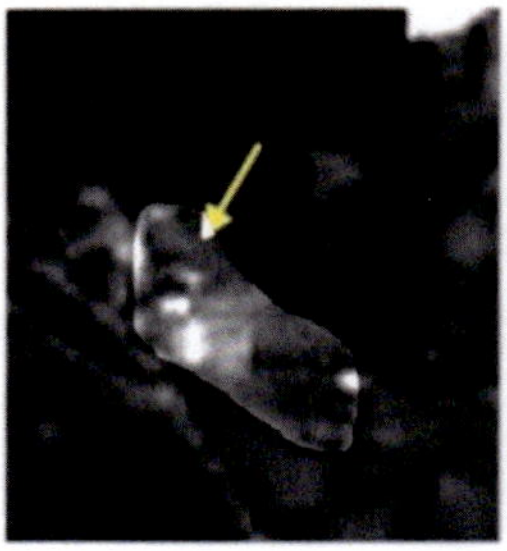

Looking Left
Time Stamp 25 sec

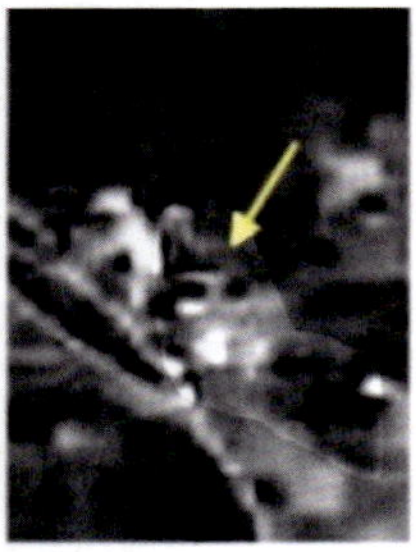

Looking forward
Time Stanp 40 sec

The next set of still captures will demonstrate the eye of the Bigfoot adjusting to the infrared light. Humans cannot see light in the infrared spectrum. It is assumed that animals like deer cannot see infrared light as well. This is why it is used on the trail cameras to illuminate the darkness because it is not detectable by the wildlife. Based on this trail camera video I think that the Bigfoot's eye is sensitive to light in the infrared spectrum and they can see the infrared light just as we see the light from a flashlight.

The still capture close ups on the next page compares the eye during the infrared flash while the iris is pinpointed and during normal light conditions and dilated.

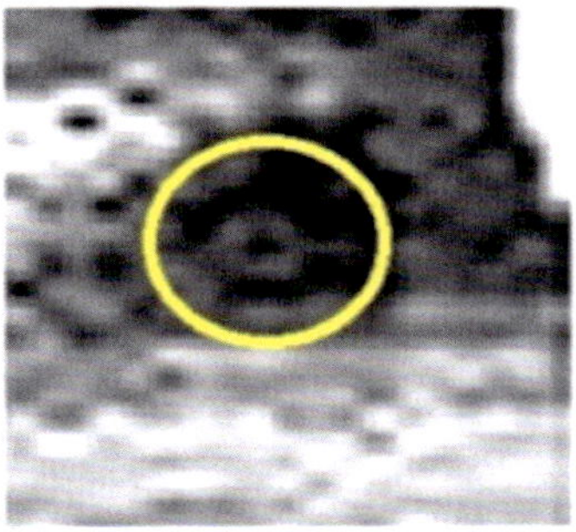

Iris pin point during flash

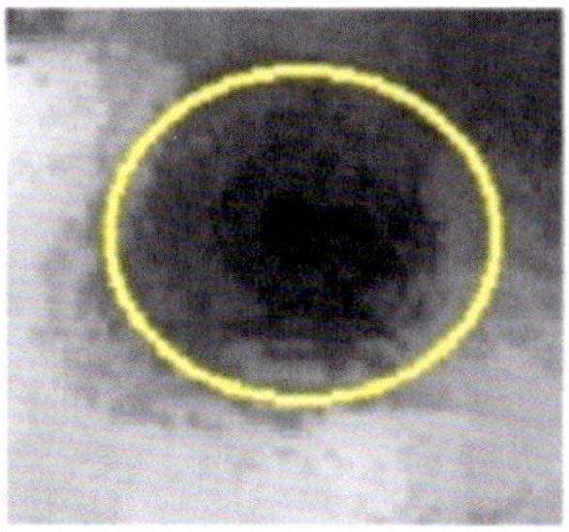

Iris dilates back to normal size
15 seconds later

A look at each eye of the Bigfoot

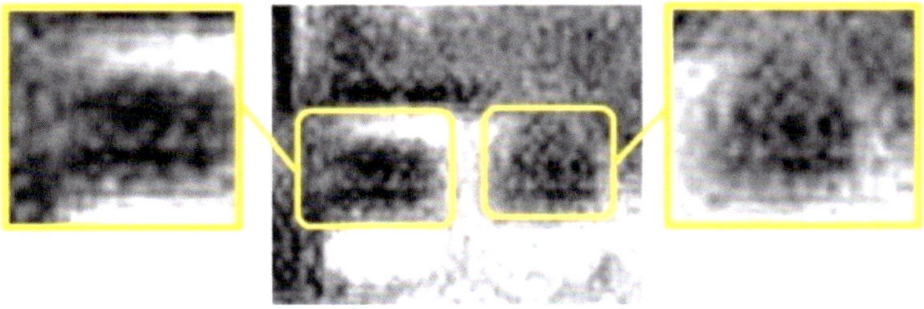

The still captures below show the left and right eye at different time stamps during the video. The iris changes sizes and adjust to the infrared light throughout the sixty second video clip.

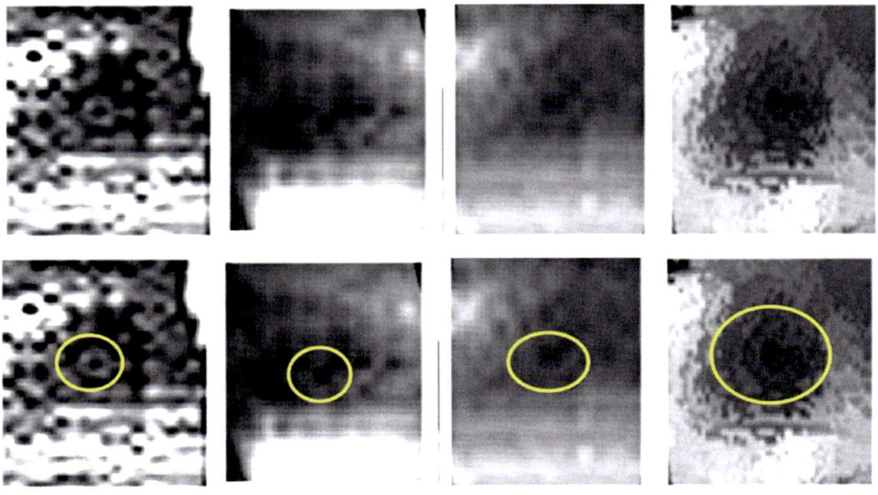

This video still receives both positive and negative comments on YouTube. I find it interesting that those who view this video and do not think it shows anything seem to resort to negative personal attacks when they comment on this video. I am fond of the saying "You get the most flak when your over the target".

The secondary camera which was placed to the right of the stick shelter was tripped but there was nothing in the video. Whatever passed in front of it was faster than the trigger speed of the camera. I had the second camera watching the trail approaching the stick shelter and it recorded nothing. After this video I began pointing the trail cameras at each other to try and catch the Bigfoot if it attempted to disturb the trail cameras again.

Chapter 13- A Family Outing, Rock Throwing, and an Experiment

Date: October 11th, 2009
Location: Research Area A

The weather was nice on this day and my wife had been asking me about my research and where I go. So, I thought it would be a good day to take everyone to my research area. The family could get some fresh air, see where I run off to on Saturdays, and could help me with a little digital camera experiment.

When we arrived at the area we went into the woods and wandered around playing and talking for a little while. The boys went down to the water's edge playing loudly and throwing rocks into the lake. I was hoping that the noise of children playing would attract the Bigfoot in my research area.

I had my video camera and while the children were playing I went to check my trail cameras and the stick shelter. My wife stayed to watch the children as I ventured off on an uneventful check of my research area. It would not be uneventful for my wife.

After I walked away my wife walked down toward the lake shore just to look out on the lake and enjoy the warm October sun. My wife was standing on the edge of the water enjoying the view when she looked right to check on the boys as they were playing along the shore. The children were approximately seventy five feet away to the right of my wife. After looking at the children she began to turn her head back toward the lake when she heard something hitting limbs as it flew through the air. She turned her head to the left in time to see a rock about the size of a baseball coming toward her from out of the woods. Before she could even react the rock hit the ground a few feet from her to the left, bounced a couple times, and then went into the lake. She spun around quickly and looked into the woods. She did not see anyone. She quickly looked toward the boys who were still playing and unaware that anything had just happen. My wife was in shock at what had just taken place. She had listened to me tell stories about the Bigfoot throwing rocks at me and other people, but could not believe it had just happen to her. Unfortunately this was the only activity she witnessed. The rest of the afternoon there was no activity around her or the boys. Below is the location where this incidence occurred.

The Camera Experiment

Many times when I am out in the woods I will get the feeling I am being watched. I will turn and film several minutes of video where I think whatever is watching me is hiding. Then when I return home and review the video I will see what looks like a face looking at me. It is usually gray or green because of the distance from me and the algorithm the camera uses to determine color. Many times I will see two eyes, a nose, and the round or dome shape of a head. These "faces" could be explained away as anomalies, and that my brain finds the "face" in these anomalies, but what if these "faces" are not anomalies?

In order to determine what a "real" person would look like in similar circumstances and distances I had my wife help me with an experiment. She took video clips of me at various distances and in different areas of the woods. When I reviewed the videos I found something very interesting. The video camera's algorithm turned my face gray or green depending on the surrounding vegetation. When I compared the videos of me with the videos of what I think are possible Bigfoot the similarities were astounding.

Below is a still shot from the test video

The above photograph demonstrates how the digital camera will change the coloration of my face from flesh tone to green. I am kneeling down and peeking over thick green vegetation. Notice how my eye sockets are very dark with no eye detail. I even seem to have a "dome" shape to my head. The photo above is a still capture from a high definition video camera. I am approximately twenty five feet from the camera.

The still capture above demonstrates how the camera can make me look like a "ghost" figure. I am approximately seventy five feet from the camera during this video. You can see how I lose all "human" appearance and my face actually resembles a bare skull. Also note again my skin coloration is light green.

Below I compare the test still captures with suspected Bigfoot still captures. In each photograph the suspected Bigfoot is on the left and the test photograph of me is on the right.

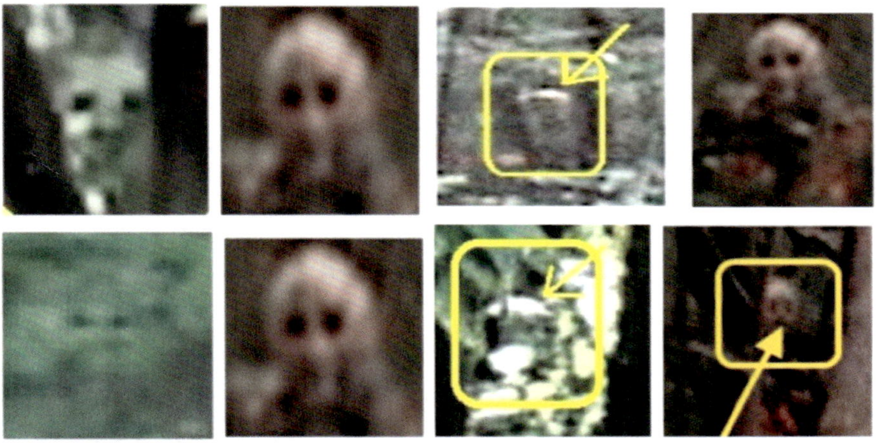

Chapter 14 - Hunting Season, New Cameras and Rock Piles

Date: October 17th and October 24th, 2009
Location: Research Area A

October 17th, 2009

The muzzle loader (Gun) hunting season had begun and the presence of men in the woods with guns had influenced the activity of the Bigfoot. When I went to the woods on October 17th there was no sign of any activity at the stick shelter or on the trail cameras. I did hear the faint sounds of footsteps that let me know I was being followed but the Bigfoot seemed more timid and stayed further away from me than normal.

I decided to add some additional things to the stick shelter in order to tempt the Bigfoot into messing with them and hopefully catching them on the trail cameras. The first thing I did was to take the larger flat rock and move it toward the back of the shelter so my trail camera would be pointing straight at it about 15 feet away. I also took some sticks and formed a small pyramid over the rocks. Some critics on YouTube had made the assertion that the rocks were being moved and rearranged by animals, by putting the sticks over the rocks I would be able to tell if animals were causing the rocks to be moved.

I also found a flat piece of concrete block and created a smiley face on it to see what would happen.

I also purchased two inexpensive High Definition (HD) video cameras to assist in my research. I found they took good quality videos if they were stationary. Since the inexpensive cameras did not have the "stabilization" feature the videos were blurry if I attempted to take video while walking. They would prove very helpful when I was working a location like the stick shelter. I could set them facing my blind sides to monitor the area for any Bigfoot sneaking up on me. I had captured a majority of my Bigfoot footage while I was looking one direction and filming in the other. I was hopeful that placing these cameras to monitor my blind spot would increase the odds of capturing a Bigfoot on HD video.

I placed my "new" cameras around me covering areas where I had captured video footage of the Bigfoot on earlier visits to the stick shelter. I then down loaded the trail cameras and gathered up the cameras I had set out.

I made my way to my hunting stand and spent the rest of the day enjoying a peaceful day in the woods. I returned home that evening and reviewed all the trail camera footage and the only activity captured on the cameras were two squirrels. The cameras I had set out to watch my blind spots did not capture any video of interest.

October 24th, 2009

When I returned a week later to the area I found some interesting things had happen to the smiley face.

Smiley face on 10/17/2009

Rocks moved 10/24/2009

I was very hopeful that the trail cameras captured what moved the rocks. I was also hopeful I would see a Bigfoot squatted down moving the rocks around.

I also noticed that the rocks I had placed under the stick pyramid were unmoved, but the bottle placed beside the stick pyramid was knocked over. I found this odd. The stick pyramid was constructed to fall apart with very little disturbance. I made it this way on purpose so I could tell if an animal lumbered through the area and knocked it down. I reasoned if the rocks underneath were moved it would be impossible for an animal to do this. The rocks were not moved but the bottle I had placed against the side of the pyramid had been knocked over. How would an animal knock the bottle over without toppling the pyramid? Did a Bigfoot disturb the bottle but leave the pyramid alone? Is it possible the Bigfoot recognized the "trap" that was set and reasoned that if it moved the rocks under the sticks I would be aware of its presence?

With all this activity, I anticipated I may have captured a Bigfoot on the trail camera. When I reviewed all the videos captured by the trail camera for that week I had one video that was black! The rest of the videos were of squirrels or deer. I tried for several hours to enhance the video in order to see what, if anything was captured on the clip. I was unable to get a viewable image. I was learning how unreliable these trail cameras could be.

I had experienced the infrared being so bright that it would white out the screen but I had not had the video be so dark you could not see what was going on. The only thing I can think of that could cause this is something was covering up the infrared emitters thus effectively blinding the camera. That begs the question, are the Bigfoot that intelligent? Did they cover up the infrared emitters? In the future I will have multiple times where something is so close to the camera that you can only see the white wash out from the infrared reflection. Do they know this is a camera? Do they realize that by placing something over the front of the camera it cannot take their picture? Or is it just instinct to cover up the square box that puts out a strange red glow?

I think the Bigfoot definitely moved the rocks around on the smiley face. What did this mean? This was the pressing question on my mind. I went to my deer stand after checking the stick shelter area and then returned home.

Chapter 15 - Bigfoot Sign Everywhere

Date: November 1st, 2009
Location: Research Area A

The day was a bright sunny day with cool brisk winds. It had rained several days the previous week so the trails were muddy and full of water. The ground was covered with freshly fallen leaves. I worked my way down the trail toward the stick structure and began to cross the small stream that only ran when there was significant rain fall. I stepped across and noticed on the side of the stream bank a large heel imprint. The leaves were still compacted into the print and muddy, so what ever made the print passed by this way in the last twenty four hours.

I worked my way down the trail to the large mud hole that I regularly have to traverse when traveling this way to the stick shelter area. I have already found prints (toe slides) in this mud hole and this day I would find a complete footprint.

My boot is thirteen inches long and five inches across at the ball of the foot. This makes the footprint at least sixteen inches long and seven inches wide at the ball of the foot as demonstrated in the below photograph.

I think this is the same Bigfoot that left the heel mark stepping across the creek bed only twenty five yards away. The size and dimensions are similar to earlier footprints I have found in this area. I think this print belongs to the large Bigfoot I have named "Black Nose".

I documented the footprints then moved to the stick shelter area. The leaves were very thick on the ground. I did not see any sign of activity on the ground, but when checking my trail cam I noticed a two muddy finger smudges on the trail camera.

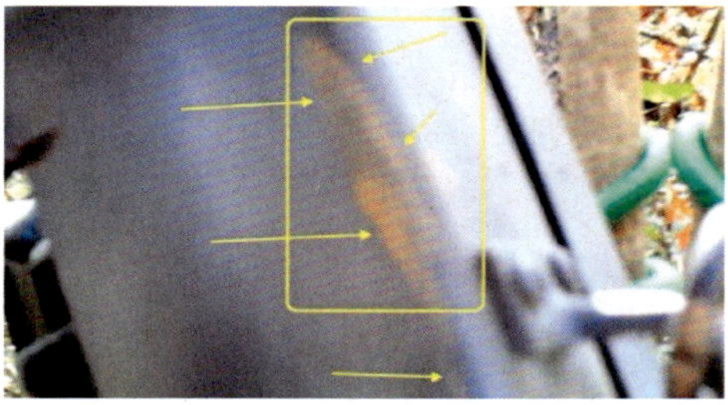

When I downloaded the video clips from the trail camera I found out why. On October 25th, 2009 at 5:37 PM a Bigfoot stood beside the camera reached out and shook it. I know this because I could see the shadow of the arm as it was raised up to shake the camera. I have tried to get a good still capture of the shadow but was unable. It is very clear when you watch the motion of the video. The Bigfoot was smart enough to know he would cast a shadow so he was standing in the shadow of a large tree, but when he reached his arm out it was in the sun and cast a distinct shadow.

After finishing up at the stick shelter location I worked my way around the long point where I had my first encounter while fishing. The trail was very grown up but as I made my way down the trail I noticed a very distinct path of limb breaks going cross the trail and down toward the lake. I followed this path and found numerous limb breaks, some at the ten to twelve foot level. The "limb break" trail made a large loop around the point then headed back toward graveyard. There were no limbs or trees on the ground around these breaks. Many times limbs just a few inches away were left intact. The limbs appear to have been broken selectively in order to mark the trail.

I followed this trail for a couple hours finding dozens of broken limbs and disturbed ground. It was obvious that this was a route that had been used many times from the age of the breaks. It resembles a "patrol" route around its territory. I wonder if they walk this route on a regular basis checking for people who may have entered this area.

The trail took me back to the cemetery where my truck was parked. I finished up and left for the day.

Chapter 16 - New Trail Cameras, Footprint, and Stick Structures

Date: November, 14th, 2009
Location: Research Area A

I purchased two additional trail cameras to augment the two I already had in the field. One of the cameras was a high end Bushnell Trophy Camera. This camera takes 5.0 MP still photos and high quality video clips. The other camera was a 3.0 MP black and white Stealth Camera. It is of lower quality but reliable and cheap. I devised a plan to deploy three cameras at the stick shelter location in a triangular pattern and the fourth watching a "mock" stick structure I planned to build.

I entered the woods at the normal location and made my way toward the stick shelter with the new trail cameras in my back pack. While going around the large mud hole in the trail near the stick shelter I notice a large footprint in the mud hole very close to where I found a previous footprint back on September 12th, 2009.

The footprint was seventeen inches long. Similar to other footprints I have found it was located on the edge of the mud hole made as the Bigfoot attempted to avoid leaving footprints. The Bigfoot has a habit of going around any area where they could possibly leave footprints or any sign of their passing. The only exception seems to be when an area is covered in water. I have found they will step in these areas or on the edges of these areas. Once the water dries up I occasionally find the remnant of a footprint like the one above. I think they are intelligent enough to know that they will not leave clear footprints behind if they step into a mud hole filled with water.
After documenting the footprint I make my way around to the stick shelter and start setting up my new trail cameras. I form a triangle with the trail cameras. I position the Bushnell camera to cover the rock arrangement. I place the black and white camera watching the Bushnell camera in case a Bigfoot tries to sneak up from behind it. I place the third trail camera watching the trail that leads over to the small island near the stick shelter.

I setup the trail camera on the trail leading to the small island then worked my way to the end of the point where I had my first experience with the Bigfoot. My plan is to make a mock stick structure and place a trail camera watching it.

I placed the cameras and made my way back out to my truck and left for the day. When I arrived home I reviewed the trail cam videos. Nothing was captured on them except for two squirrels playing.

Chapter 17 - Bigfoot Hiding in the Brush

Date: November 25th, 2009
Location: Research Area A

Deer season was in full swing as I entered the woods this morning. I am an avid deer hunter and had put the Bigfoot research on the back burner as I concentrated on deer hunting. This day I had decided to hunt in the morning then do a little research and check my trail cameras in the afternoon.

I made my way into the woods approximately thirty minutes before day light. Almost immediately I began hearing the familiar sound of bipedal walking to my rear. I continued on my way intent on going hunting and allow the Bigfoot following if it wanted to. This day was different than others because instead of staying back and following me the Bigfoot came closer and was making a great deal of noise. It was purposely breaking limbs and almost stomping at times as I continued to walk. It was obvious the Bigfoot wanted my attention or wanted me to stop.

I was nearing the area I wanted to hunt deer and had to get this Bigfoot to leave me alone. This loud display the Bigfoot was putting on plus its smell would keep any deer in the area away from me. The Bigfoot kept back about fifty yards, half the usual distance they usually stay back when following me. Then I heard a rock hit the trail in front of me. I took another step and again a rock came from my right and hit in the trail directly in front of me. I stopped, turned, and looked in the direction that I heard the noises. I said the following out loud "Hey, I am trying to go hunting and you are making way too much noise. If you want to come along that is fine, just quiet down!" I then turned and continued to my deer stand. The Bigfoot must have understood the message because it dropped back to approximately one hundred yards and quieted down. It still followed me and watched me during my entire hunt, but at least it was quiet. The Bigfoot did position itself up wind of me and I smelt it during the entire hunt. Any deer in the woods around me did also. The Bigfoot knew that I was deer hunting and it was just frustrating to have my hunt disturbed. This is one of the unintentional consequences of my situation and if I wanted to do any serious deer hunting it would have to be in a different area.

I endured the morning hunt seeing no sign of deer or any other wild life for that matter. The presence of a Bigfoot that close to my stand acted like a "repellant" to all large and small game in the area. I decided it was time to give up hunting and go check the trail cameras. I started walking the horse trail toward the stick structure location when I started hearing the bipedal walking again. This time instead of behind me the Bigfoot was staying out ahead of me as I walked. I kept getting the strong feeling I was being watched.

This experience was different. The Bigfoot did not follow me from behind like they had in the past or earlier that day. The Bigfoot was now out in front of me. I think there was more than one Bigfoot watching me walk down the trail. It appeared to me that they were taking turns watching me. One would get ahead of me on the side of the trail, hide, wait for me to walk up the trail, and then watch me. After I would pass the Bigfoot would get up and run ahead of me again and setup. While one was watching the other one was moving ahead of me and getting into position.

I approached an area where the vegetation was very thick on both sides of the trail when I was hit with infrasound. The hair on the back of my neck stood up and the eerie feeling I was being watched was very strong. I stopped and looked to the right of the trail first. Below is the area where I was looking on the right side of the trail.

Then I heard some movement directly behind me so I slowly turn around and start looking in the opposite direct. While I was turning the camera caught two frames of a Bigfoot that was hiding in cedar tree. I think at least two Bigfoot were around me, one off to the right and one off to the left. The Bigfoot on the left was attempting to get my attention so the one on my right could sneak off.

I was experiencing infrasound, but it was different from what I had experienced in the past. I was not having the feelings of dread and high anxiety. It was almost like they were trying to get my attention, wanting me to look in their direction. I felt like I was being studied and watched. I also think the Bigfoot on the left side of the trail was trying to get me to walk toward it so the Bigfoot hiding about twenty feet away could sneak off without me seeing it, but I stayed put and just zoomed in with my camera.

I realized what was going on and turned around and started looking on the right side of the trail again. I knew the Bigfoot were there, I just could not see them. So I slowly panned back and forth with my camera hoping even If I did not see them the camera could possibly capture what I could not see on film.

When I was reviewing this video my suspicions were confirmed and I had captured video of a Bigfoot hiding in a unique cedar tree. I had zoomed in and was slowly panning right to left when I captured the following.

The above still capture shows the Bigfoot perched in the cedar tree

I had captured seventy five frames of this Bigfoot. Most of the upper body is visible along with the left arm, left hand, and fingers.

In one of the still captures from the video you can see detail of the left eye of the Bigfoot

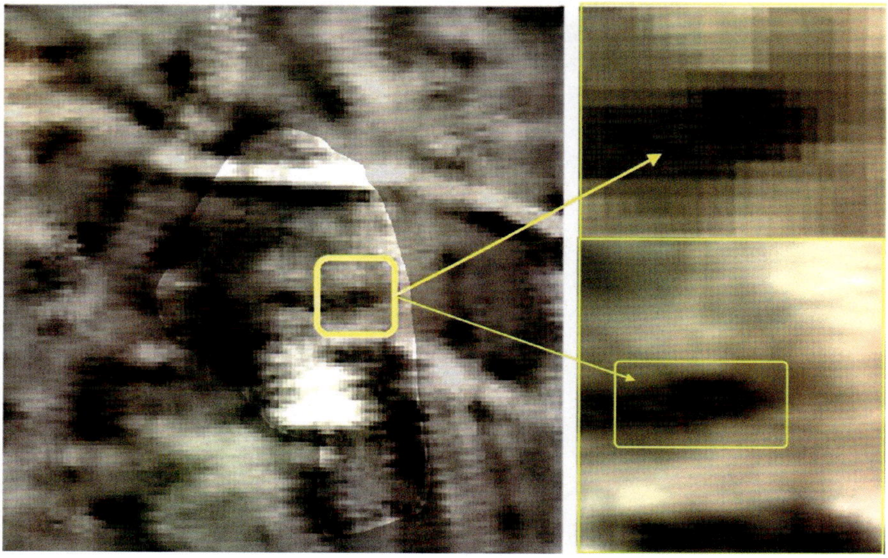

There are so many elements to the video. In my opinion the head, shoulder, arm, hand, and foot are visible. The way the Bigfoot is perched in the cedar tree its foot and toes are visible. In the below still capture the cedar tree is outlined. The hand and foot is highlighted in a yellow square.

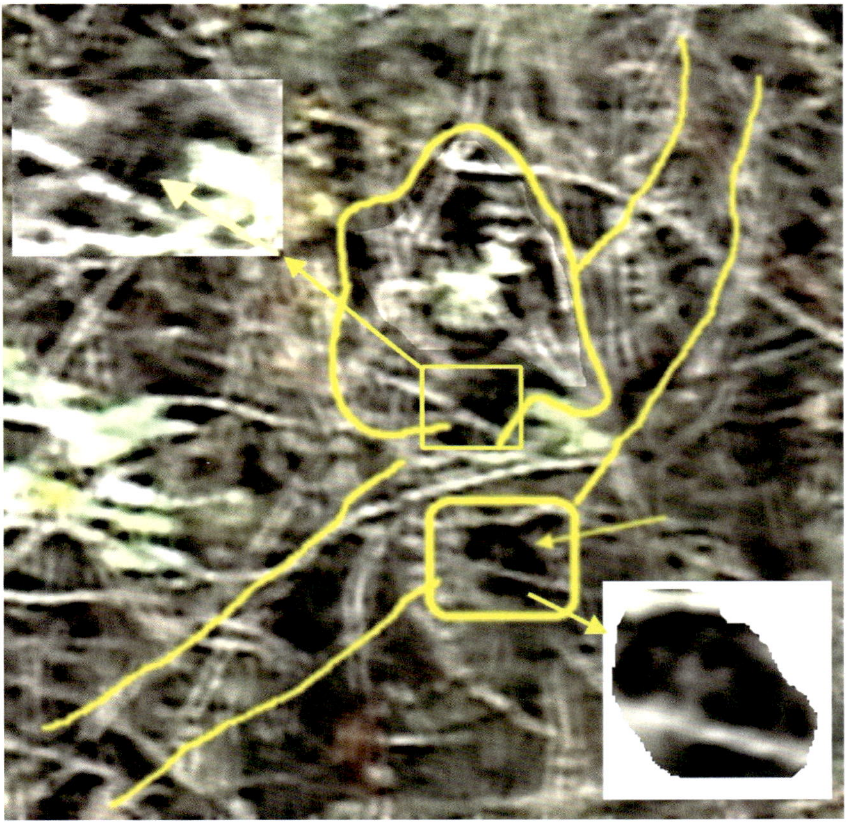

In the weeks to come I do extensive follow-up on this video. I even place a trail camera in area watching the cedar tree. I was hoping that the Bigfoot used this tree as a look out location to watch people and the horse riders as they ventured down the trail.

I found an abundance of trace evidence in the area where the Bigfoot was sitting in the tree. This trace evidence included dozens of small broken limbs, wear marks on the bark of the tree, and freshly stripped bark on larger limbs that are consistent with a large size hand.

When I stood in the location where the Bigfoot was hiding I noticed the bark was freshly twisted off a limb. The size of the damage was five inches long. I think this is where the Bigfoot was holding on with its hand.

The following is a facial collage of the Bigfoot at different points during the seventy five frames I was able to capture on video. I point out the head, brow, eyes, and nose on each still capture.

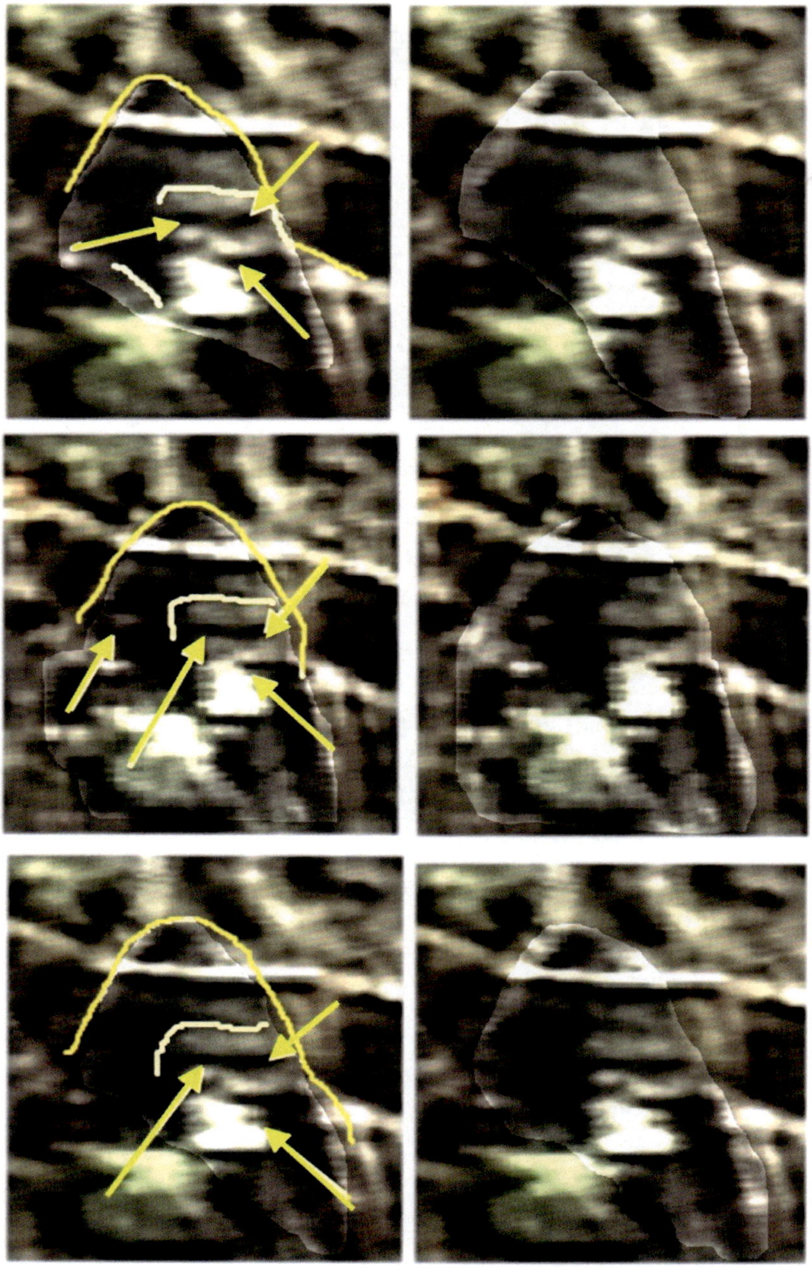

The enhancement to actual photograph

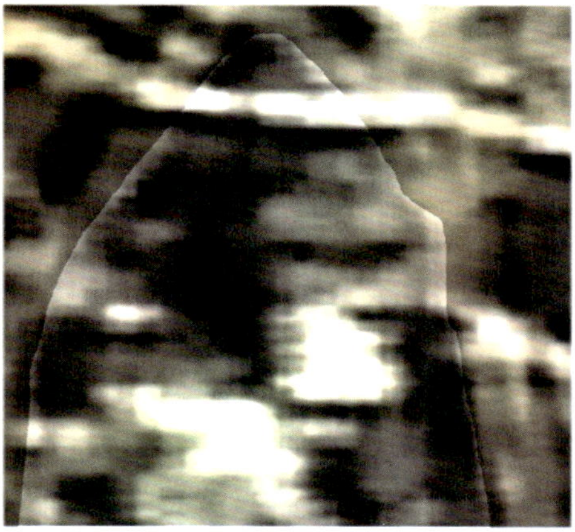

I never did see this Bigfoot when I was in the field filming. I am always amazed at how close they get to you without you being able to see them. Their camouflage is flawless and their methods of hiding long practiced and perfected. I work every time I go into the field to train my eye to spot them, to see the dome shaped head, the feet, the hand, or just one eye peeking from behind a leaf. I can "hear their song" (infrasound) for a lack of a better description, but only vaguely. I usually know when they are close to me but where they are or where they are going is still a mystery.

I leave the area and continue on to the stick structure and download the videos from the trail cameras. When I review the trail camera footage at home I find I have captured an infant Bigfoot feeding under one of the trail cameras. The details are in the following chapter.

Chapter 18 - Trail Camera Footage

Date(s): November 15th 2009 – December 06th, 2009
Location: Research Area A

I had arranged my trail cameras at the Stick Shelter location with my best camera, a Bushnell Trophy Camera, aimed at the rocks and bottles that usually get moved around. I placed an older black and white Stealth Camera at a right angle to the Bushnell in case a Bigfoot tried to mess with it.

I did not realize at the time that the Stealth Camera was aimed to the left of the Bushnell, so if a Bigfoot came in from behind it I would not see it. This "mistake" would prove to be very fortunate. If not for my bad aim I would not have captured an infant Bigfoot feeding under the line of sight of the Bushnell Trail Camera.

This new configuration captured the attention of the Bigfoot the day I put them up. Within approximately thirty minutes of me leaving the area something shook the Bushnell from behind. Then the next day it was grabbed from behind again. The camera did capture one frame of what looks like a "hairy finger".

Eight days later on November 23rd, 2009 a light rain was falling shortly after daylight when an infant Bigfoot makes its way to the Stick Shelter. I think it can hear the Bushnell Camera's motion sensor and stays out of range of the camera. It however does not notice the Stealth Camera mounted fifteen feet away. The infant Bigfoot finds something on the ground, an apple I left behind or maybe a candy bar and begins to eat it. The trail camera is tripped and begins recording. The rain is running down the front of the camera and causes some slight distortion but the Bigfoot can be seen in profile feeding. Then suddenly right at the end of the ninety second video clip something startles the Bigfoot. It pops its head up in a fraction of a second. The trail camera was recoding at thirty frames per second. It took the Bigfoot two frames to go from a bent over position to an upright position!

There are not many eye witness accounts of an infant Bigfoot. They have been described as looking very monkey like but with a "human" like nose. The infant Bigfoot I saw as a child looked like monkeys from the forehead up but from the eyes down the face was human like. The nose was a human looking nose and the eyes were big and dark. The mouths also looked like a humans.

The infant Bigfoot continues to feed for approximately eighty seconds of the ninety second video, and then rises up with amazing speed. There is only ten seconds of video where the face is visible but these ten seconds reveal what an infant Bigfoot looks like. In my opinion there is no conjecture or guessing required.

I took the clearest frame and after using a standard noise filter, adjusting contrast, and tweaking the RGB Levels I was able to bring out the below details.

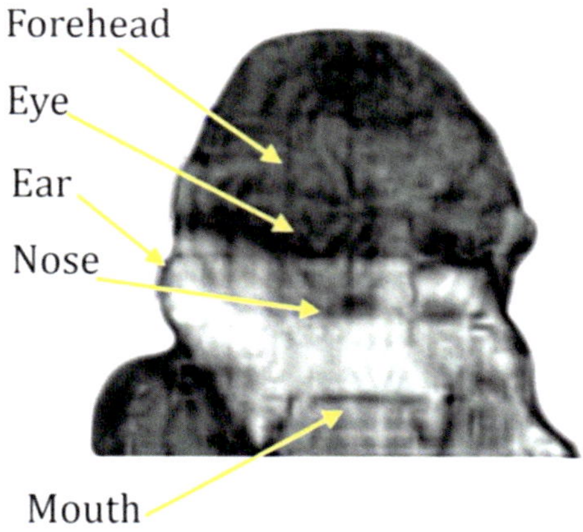

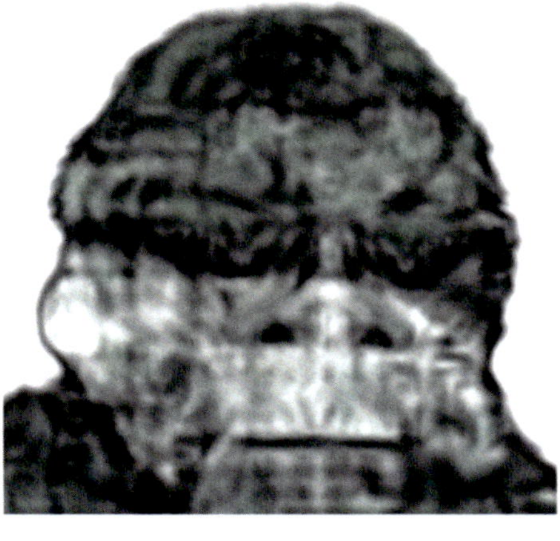

I cleaned up the eyes, nose, and mouth area. The effort produced a clear photo of the Infant Bigfoot. This was a very eventful two weeks. I had captured one Bigfoot on video and the other on a trail camera.

Chapter 19 - The Bigfoot go into "Stealth" Mode

Date(s): December 06th, 2009 to January 16th, 2010
Location: Research Area A

After a very eventful November 2009 the Bigfoot go into what I call "stealth mode". The overt activity dwindled to almost nothing but I continued to feel I was being watched. The woods were oddly quiet during this "stealth" period that lasted approximately a month. Every time I entered the woods I could "feel" their eyes on me. I was followed everywhere and the Bigfoot would make just enough noise to let me know they were watching. I wonder if their infatuation with me was over and now they were sizing me up as friend or foe? I was afraid as winter set in and the crops were harvested that the Bigfoot might move back toward the mountains. The stealth mode could also have been a result of the lack of foliage and cover. All the leaves were now down and the visibility through the woods had increased significantly. Either way, for whatever reason, the Bigfoot were now hiding and watching.

On December 6th, 2009 I made my way to the stick shelter with my Bigfoot escort in tow. It is making just enough noise to let me know that I was being followed. I continue on toward my destination stopping to check my back trail periodically to let the Bigfoot know I am aware it is following me. Once I arrive at the stick shelter I check the rocks and find no activity. I get my laptop out and begin downloading the cameras. I have not purchased extra SD cards at this time so I must down load the cards in the field to my computer then erase the SD cards and reset the trail cameras.

I continued investigating the research area but there was no other Bigfoot activity. I did find an interesting scat sample near where the Bigfoot was hiding in the cedar tree back on November 15th, 2009. It looked like human scat, but it was larger. I should have taken a sample to check for DNA but it did not cross my mind. I finished up and left the research area for the day.

December 12th, 2009 and January 16th, 2010

The holidays were now upon me and I was not able to get in the woods very much. I did make a quick trip to the research area and I placed black thread across Bigfoot trails in several locations. I also placed black thread across the entrance of the hide near the lake shore and around the stick shelter. The thread was placed at the seven foot level so if a person or animal wondered underneath they would not break it. The purpose of the thread is to detect Bigfoot movement along these trails.

After we made it through Christmas, New Years, and a bout of bad weather I was able to return to the research area on January 16th, 2010 for a full day. I did some brain storming during my down time and decided to rearrange my trail cameras and try some new experiments.

The first place I visited was the hide or blind near the lake shore. I have previously documented this area in Chapter 2. I had placed black thread across the back entrance to the blind at approximately six feet high between two trees. When I checked the thread it had been broken!

Though I cannot say a Bigfoot broke the thread going into the "Blind", I think it is a very strong possibility since the thread was at six feet. This is another one of the coincidences that regularly occur in Research Area A.

After documenting the thread break I tie a new strand across the opening and work my way to the stick shelter to check on the rocks and rearrange the trail cameras. When I arrive at the stick shelter nothing has been disturbed so I download the trail cameras and then move them to a new location. I put the Bushnell® Trail back into the brush across from the stick shelter. Then I put the black and white trail camera on a tree facing the Bushnell Trail Camera in case something tries to mess with either camera the other one will be ready to catch it.

I also rearranged the rocks moving the smiley face to the back of the stick shelter and putting the flat rock back beside the tree. I also stacked some rocks up to see what would happen.

After I set the cameras I move onto the point where I had constructed a mock stick structure to see if it would could illicit a reaction from the Bigfoot. Once I arrived I noticed a large X formed with limbs that were approximately ten feet to the right of the stick structure I had constructed. Placed in the middle of the X and perfectly balanced was sapling. This made a star or asterisk looking "stick structure." The "stick structure" was constructed out of the trail camera's field of view.

I examine the stick structure and notice how intricate the "weave" of sticks is in the middle. I also find it very interesting how the sapling that is perfectly balanced in the middle was placed under the other sticks that form the star.

In the below photograph you can see how the sticks were placed into the Y of the tree to form the star

In the below photograph you can see the sapling that is perfectly balanced in the middle

This will not be the first stick structure I find in Research Area A. This "Star" pattern has a meaning to the Bigfoot. I must try and figure out what it means. I also find it interesting that the Bigfoot placed its stick structure right next to mine. Why? What was it trying to say? Was it a warning? Was it acceptance of my presence? There are so many unanswered questions.

I finished down loading the camera and decided to just leave everything alone to see if anything else would happen. I went to the fourth trail camera that I had setup watching the Cedar tree where the Bigfoot was hiding and watching me. (Chapter 17). The rest of the day went without incident and I left the area.

Chapter 20 - A Message in the Rocks, Bigfoot on The Hill, Stick Structure Experiment, Gifts and Creepy Trail Camera Images

Date: January 23rd, 2010
Location: Research Area A

I had time to do some thinking about the year of research that was ahead of me. I was gaining experience and learning about the Bigfoot in my area. I had found stick structures during the past year and wanted to use these to see if I could get a reaction from of the Bigfoot. Stick Structures are difficult to interpret. The Bigfoot builds them but what do they mean and what is their purpose? I decided to make my own stick structure and leave a gift. I did some research and this leads me to leave tobacco as a gift. It is my belief that the Bigfoot still hold to Native American traditions. Tobacco was considered a valuable gift in the Cherokee culture. With this knowledge I decided to make a mock stick structure, place the tobacco in a blue colored cloth bag, and then place it on the mock stick structure.

It was a warm day for January, the sun was bright, and the sky was deep blue. I arrived and as I was getting my back pack on I "felt" eyes upon me. I knew this was going to be a good day because the Bigfoot were watching the area where I park my vehicle. The Bigfoot began following me as soon as I walked into the woods. I took my time, walked slowly, and panned my camera in three hundred and sixty degree circles. I also would point my camera in areas where I had the "feeling" the Bigfoot might be hiding. I think I am becoming sensitive to the infrasound the Bigfoot use to communicate and this helps me home in on the areas where they could be hiding.

I make my way slowly to the Stick Shelter area in order to check my trail cameras and see if there has been any activity with the rocks I left on my last visit. When I arrive I immediately notice that the "Smiley" face has been disturbed. I had originally made the smiley face with seven rocks. When I return two rocks were missing and the remaining rocks are arranged the familiar 4 – 1 pattern or is this Syllabics?

Syllabic is a consonant-based alphabet used to write many Native American languages. These languages include Cree, Black Foot, and Cherokee.

Below are examples of Syllabics symbols

I now think the Bigfoot at Research Area A attempted to communicate with me using Native American Syllabics. On January 16th, 2010 I placed a smiley face made out of rocks at the "Stick Shelter" location. When I return seven days later they have been re-arranged.

01/16/2010 **01/23/2010**

I first thought that this was the regular thing the Bigfoot had been doing which was placing the rocks in a 4-1 pattern. I looked at this and thought what if they are trying to send me a message? What if this is Syllabics and these are Syllabics symbols? I have heard more than one interview with Native Americans in which they claim the Bigfoot knows Syllabics and they use it to communicate with the tribes.

I researched Cree and Cherokee Syllabics and found out some very interesting things. The C and the small circle ° are Syllabics symbols. Read from left to right the "word" that is formed is °C.

I then started checking the phonetics for this possible word and I found that these symbols could represent one of the following sounds in Cree Syllabics: Va Ta, Wa Ta ,Va Da, Wa Da.

Then I crossed referenced this with several Native American Languages including Cheyenne, Cherokee, Cree, Blackfoot, and the list of words from Mary Green's Book, *"50 Years with Bigfoot: Tennessee Chronicles of Co-Existence"*

I found something extremely interesting. "Wa" can be used to mean "come here", "come and see", or "visit". "Ta" depending on the inflection can mean "brother" or "friend". Put this together and the possible messages left for me were one of the following: **visit friend/brother, come here friend/brother, come visit friend/brother**.

This is definitely food for thought and the Bigfoot could have just been doodling with the rocks, but I do not think they do anything without a purpose behind it.

After documenting the rocks I down loaded the videos from the trail cameras in hopes that "something" would be captured. The Bigfoot has become very adept at avoiding them or covering the cameras so they could move around without being filmed. The only activity the cameras captured is one of the creepiest trail camera videos I have ever viewed. I actually sent this off to a well-known Bigfoot researcher and his "expert" claimed the object in the video was an owl. Now it may be an owl but it is a very creepy owl. The camera was set to take one picture a second so I do not have video action. The object moves several times and is on the left side of the frame. I think that instead of an owl or a single object, things are being held up in front camera to see what will happen to it. The Bigfoot can see the infrared light from the trail camera and I think they are trying to determine if this "object" is dangerous. The Bigfoot do the logical thing, they hold objects in front of it to make sure it is not dangerous.

That is why I do not think the object captured on the trail camera is an owl. I think this is a series of objects placed in front of the camera to make sure it was safe. Then at the end of the video clip a Bigfoot (or whatever it is) cautiously leans out with just one eye and looks at the trail camera. It is only a theory, but I think very plausible one. The Bigfoot are not "dumb animals". I think they are another race of humans. They have displayed remarkable intelligence, problem solving abilities, personality, and curiosity.

Due to limited space I am only going to show the odd still captures of what appears to be some sort of entity peering at the camera with just one eye. The entities eye looks "large and bulbous" and its cheek is pronounced. I do not see any hair on the face or the head. To be blunt this is just creepy!

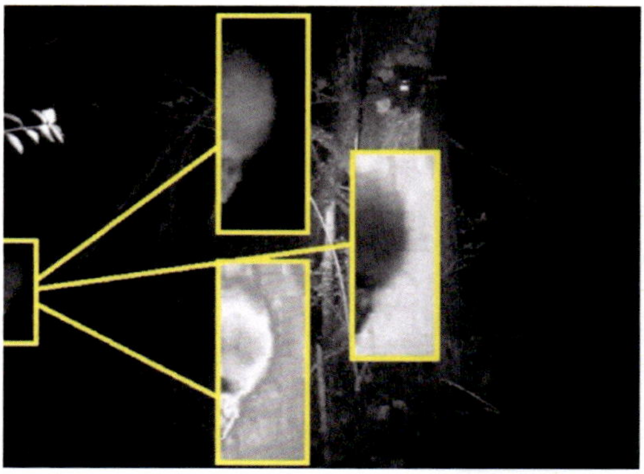

The above photograph shows what appears to be the profile of some sort of humanoid. The question now is how big is the subject?

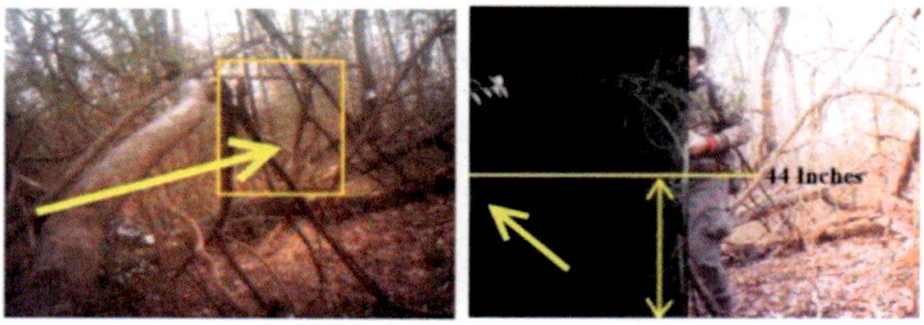

In the previous photograph the left frame is a view of where the subject was from the trail camera's view. The photograph on the right is a height comparison. Both photographs are from the same trail camera that took the still captures and in the same location. I estimate that the subjects "head" was approximately forty four inches off the ground.

In the above photograph on the left I am comparing my head size the subjects head. The scale is accurate so the subjects head is approximately ten to eleven inches chin to top of the head. My head is nine inches from chin to the top of my head. This would be a giant owl. Is this a Bigfoot? Is this something else? On the far right is the profile from the video. The head is large, but is this caused by reflection of the infrared or does the subject have a large oversized head?

I also worked on a composite photograph of the subject and have included it below. I have no explanation for what this "thing" could be. I will let you be the judge.

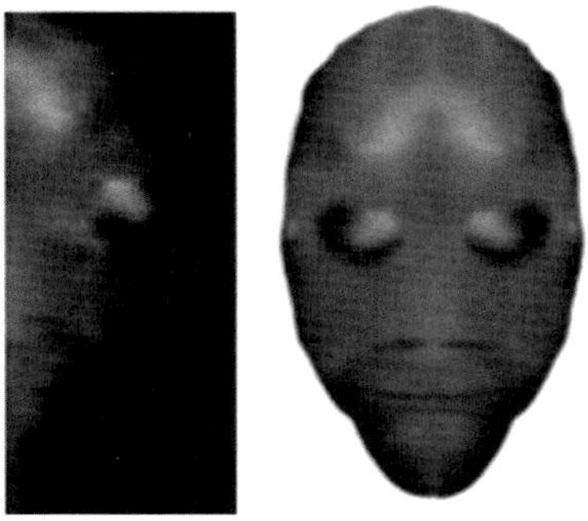

This video has been a mystery to me ever since it was captured. In my opinion this is a small Bigfoot looking at the trail camera. It was trying to figure out what the camera was and if it was dangerous. Of course I cannot prove this but I am open to suggestions. Everyone I have shown this video to is left scratching their heads and wondering.

After downloading the trail cameras I reset the rocks and the cameras and moved on to investigate the point area. I had decided to create a mock stick structure and leave a gift for the Bigfoot in this area. I started on my way to this area. In the below aerial photograph I have marked the area where I plan to build my mock stick structure. Earlier in the year I located a stick structure in this location. I think it was made by the Bigfoot so I thought this area would be a good choice.

I made my way to the old road that leads out to this area. During the entire way I had the very strong feeling I was being followed and watched. I kept hearing the sound of something trailing behind me approximately one hundred yards back. I kept stopping and filming areas where I thought I was being watched from.

I was working my way toward the mock stick structure when I was overwhelmed by the feeling I was being watched. The feeling was so intense I knew where it was coming from! The below photograph shows the area where I felt the Bigfoot was before I zoomed in with my video camera.

Once I zoom all the way in I can see the Bigfoot laying on the ground looking at me. I can see the deep set eyes and the light reddish-brown crest of its head.

This Bigfoot has all the familiar characteristics, the hair is reddish-brown, it has a pronounced brow, deep set eyes, and the skin appears to be a tan or light gray in color.

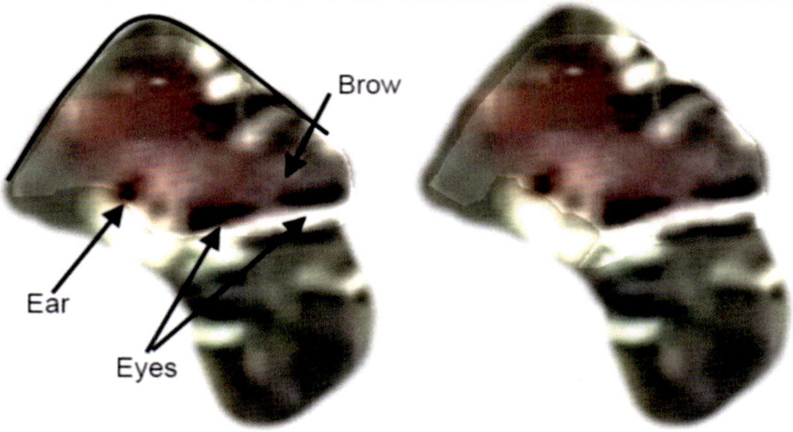

This Bigfoot looks very similar to the Bigfoot that was watching me download my trail camera back on May 30th, 2009. I think this could be the same individual in both videos. The deep set eyes, hair color, and location of ear on the head are similar.

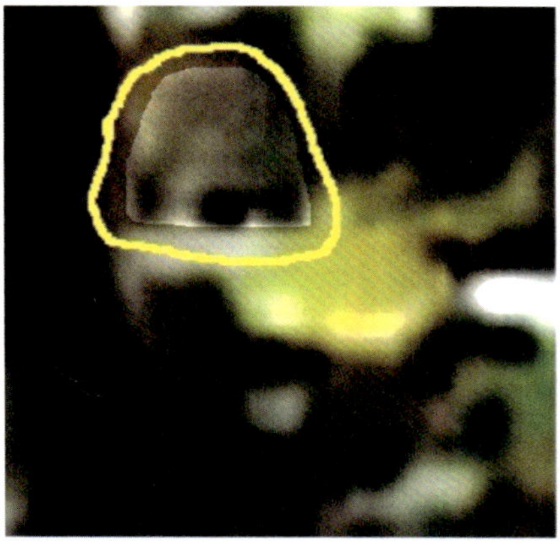

05/30/2009

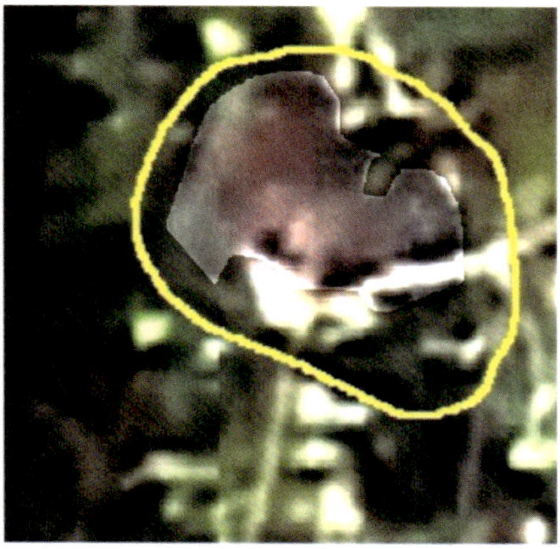

01/23/2010

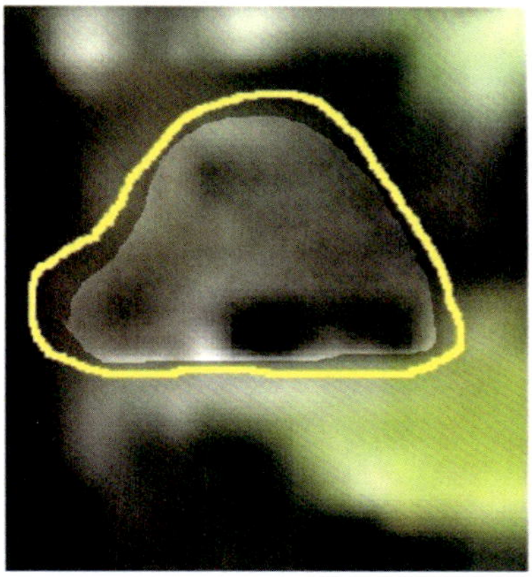

05/30/2009

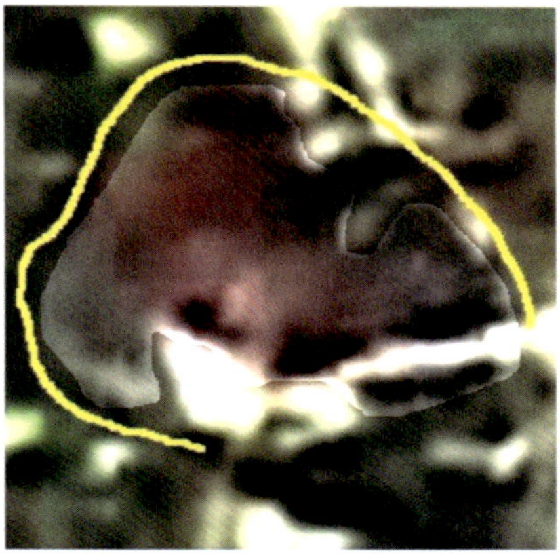

01/23/2010

I make my way up the hill toward the area where the Bigfoot was hiding as fast as I could. It took me about five minutes to get into the area. The Bigfoot was gone by the time I arrived. I did not expect him to stick around and wait on me. I began looking around for any sign. I spent over an hour in this area. I found a faint trail that led off into the thick brush so I decided to follow it. Approximately twenty five yards down the trail through the thick brush the area opened up and there was a clearing right in the middle of the heavy brush and vegetation. I find two very fresh limb breaks at a height of seven feet as I following the trail.

This would be the perfect place to hide and live. It is right in the middle of some of the thickest brush in the area. The trail entering this area was very faint and difficult to travel.

I could not find the Bigfoot and to be honest did not expect that I would. I knew it would leave the area before I could get close, but I was hoping maybe it would interact with me somehow. I think the fact that I zeroed in on its location then headed right toward it had it spooked.

I found my way back out and came to the area where I was going to leave my gift on the stick structure. I had not visited this area since I found it back in August of 2009. It has now been five months and I was very surprised to find that all the trees that formed the stick structure had been pushed down.

The most interesting thing about the stick structure was how the saplings were "weaved" in the Y notch live sapling.

When I arrived at this location I expected the structure to be intact, instead it was all on the ground. All the saplings were facing the same direction. There were no limbs or fallen trees near that could have caused all these saplings in this clearing to have fallen over in the same direction.

01/23/2010 **05/30/2009**

The previous photograph on the left shows all the trees that made the stick structure on the ground. All of them were knocked down in the same direction. All the damage is isolated to this area. Did the Bigfoot observe me around the stick structure back in May of 2009 and push down the stick structure?

I begin looking for a small tree with a Y to make my own stick structure. In this structure I will leave my gift of tobacco in a blue colored cloth bag. I find a tree just a few feet away from the original Structure. The fork in the tree will hold two limbs that form the X. Then I balance a limb in the center to form a "star". I have seen this structure in another location in my research area and also in other researcher's videos in Texas and Washington.

Will my gift be accepted? Does this layout mean anything to the Bigfoot? If so what reaction will I receive to it? All these questions would be answered in the weeks to come.

I also started playing Native American music, chants, and jump dances while I was working in the area. I think this could draw in a Bigfoot to my location given the close relationship between the Bigfoot and Native Americans.
I made sure I had several Cherokee selections since I am in the area of Chota, the ancient capital of the Cherokee.

The stick structure finishes up my outing for the day and I make my way out of the area without incident.

Chapter 21 - Minimal Activity – Rock Movement

Date: February 7th, 2010
Location: Research Area A

I started my outing in the early afternoon this day. Parking in the normal location I entered the woods and within five minutes had my Bigfoot "escort" following me. This is now a common occurrence that I have become accustomed to and to be honest I now expect to be followed. The Bigfoot station scout(s) watching the area and when someone enters the woods they are followed and watched. Since I had become familiar to the Bigfoot they appear to be more nonchalant when following me. The one following today made a great deal of noise which is uncommon for the Bigfoot. I think they wanted me to know they were following because when a Bigfoot goes into "stealth mode" they can be silent, so I knew the noise was not made by accident. They always stay approximately one hundred feet behind me, but never directly behind me. The Bigfoot follow me on one side or the other depending on the thickness of the vegetation they have to navigate and wind direction. If I walk fast down the trail it appeared to annoy the Bigfoot. I think it annoyed them because they would have to walk through thick vegetation to keep up with me. I would get rocks thrown at me and would hear loud limb breaks when I would be moving "too fast". It was the only thing I could do to appear to perplex them. Normally the Bigfoot would torture me by playing their cat and mouse games.

I worked my way to the stick shelter area and immediately noticed the "Smiley Face" had again been changed as I walked up to the shelter.

I noticed that the rocks had been moved into a pattern very similar to what I had found on January 23rd, 2010. Did the Bigfoot leave me another message in Syllabics? In the below comparison I show how I found the rocks on January 23rd, 2011 and how I found them on February 07th, 2011. The pattern the rocks are in is not exactly the same but very similar.

The week before on January 23rd, I theorized the Bigfoot left me a message using Syllabics. The rocks formed the Syllabics symbols "°C". This formed the word "Wa-Ta" which loosely translates "visit friend". Staying with this logic I looked for possible Syllabics symbols in how the rocks were arranged on February 2nd.

The below diagram shows the word “wa-pa-wa” formed using Syllabics.

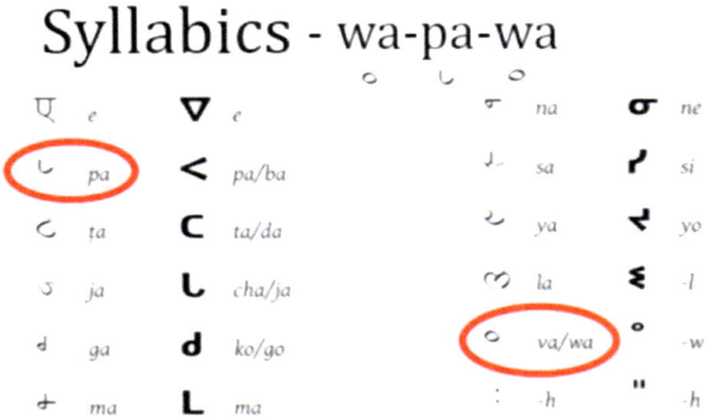

I searched several Native American language databases but was unable to find the word “wa-pa-wa” I may have mistranslated the Syllabics or the Bigfoot just moved the rocks without the intention of leaving a message.

I also notice that two small limbs directly behind the smiley faces had been broken since my last visit on January 23rd, 2010. After documenting the smiley face I set up a second camera to watch the brush to my blind side as I worked around the structure.

I have captured most of the Bigfoot pictures while pointing the camera in the opposite direction I was looking so I decided to purchase an inexpensive HD camera for this purpose. I then checked my trail camera and downloaded the video. The camera did look like it had been moved slightly but It was hard to ascertain by what.

I reset the rocks again, downloaded the trail cameras, and then moved on down the trail toward the mock stick structure I had previously built on January 23rd, 2011. I was very anxious to see if there had been any activity around the stick structure. I was hoping my gift of tobacco had been taken. Unfortunately when I arrived I noticed there had been no change in the area. Nothing had been altered or taken. I was disappointed but I knew this was something new from me and it may take the Bigfoot some time to get used to this change in my behavior. I documented the area then headed back out to the truck and left the area for the day.

Chapter 22 - Rocks, Stick Structures, and Bigfoot Games

Date: February, 20th, 2010
Location: Research Area A

This trip would be an interesting one. The Bigfoot would follow me all day and let me know that they were all around me. From time to time they would break small limbs and make loud footfalls. When I entered the woods I was picked up almost immediately and followed to the stick shelter location where I would check the trail cameras and see if anything had moved the rocks around. I also had my new HD video cameras with me. They are inexpensive and do not have the stabilizing capabilities that the more expensive cameras have, but for capturing video from a fixed position they will work well. Since most of my videos that capture a Bigfoot are filming my blind side, I decided to place cameras covering the entire area. This will hopefully capture a Bigfoot if it decides to sneak up from behind me and take a peek at me while I have my back to them.

After setting my cameras out I begin my work to down load the trail cameras and document any rock movement. I go to check my number one trail camera and to my disappointment I notice it has malfunctioned. I think the batteries have gone dead, but I will not know until I return home and check. I also notice that the colored rocks I placed on the flat rock and been moved.

In the below comparison the rocks have definitely been moved since I left them on February 7th, 2010.

Setup on 02/07/2010

Return on 02/20/2010

The rocks that were "moved" appear to be placed in a pattern. I find it hard to accept that a "small animal" moved the rocks or knocked them off the flat rock. The direction and the way the rocks are sitting denote to me that they were placed not randomly knocked off. Does this pattern mean anything?

This movement of the rocks increased my disappointment that the trail camera had failed! I decide to play some Native American music and chants while I work around the area. I am hoping this will attract a Bigfoot and my cameras will capture him.

I begin to download the trail cameras when the "Bigfoot Games" begin. This is not the first time this has happen to me, so I know what to expect. I begin hearing small limbs break approximately fifty yards away in the thick brush, so I turn with my video camera and watch the area intently. Three or four minutes later from the opposite direction I hear another small limb snap. So I turn and look in that direction, again in a few minutes I hear what sounds like a rock falling through the woods to my right so I turn and look in that direction. This goes on the entire time I am at the stick shelter location. It is the game the Bigfoot and I play.

I notice that there is one rock out of place on the "Smiley Face". I do not think much of this, but when I get home and review the trail camera video it will be very important.

Below is a comparison between how I left the smiley face and how I found it two weeks later.

Smiley Setup 02/02/2010 **Smiley on 02/20/2010**

I notice that two rocks have been removed and replaced with two different rocks! In the below photograph the rocks on the left were left as part of the smiley face on February 2nd. When I returned on February 20th, these two rocks have been replaced with two different rocks.

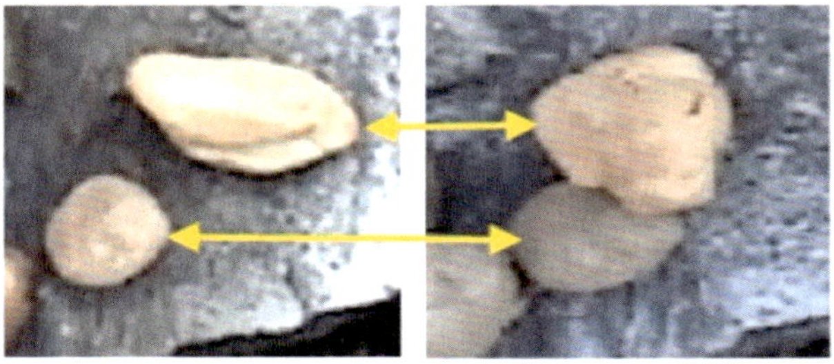

February 2nd February 20th

After documenting the smiley face I continue to play the games with the Bigfoot for a while never seeing anything. I rearrange my trail cameras since I will only have two in this area now. I also hang some strawberry flavored hard candy out to hopefully lure the Bigfoot in front of my trail cameras.

After hanging the candy I set the trail cameras and head out toward the location where I have a trail camera watching the hiding spot where I captured a Bigfoot on video in November of 2009.

The Mock Stick Structure and Tobacco Gift

After I finished down loading the trail cameras at the Stick Shelter location I made my way to the area where I had captured the Bigfoot on video in November of 2009. I have placed a trail camera watching the cedar tree the Bigfoot was perched in. I was thinking it might return and use this location to watch the horse trail and observe the people and horses as they went down the trail.

The trail camera did not capture any videos so I moved down the trail to the mock stick structure/gift location. I began to hear the faint sounds of bipedal walking behind me at the distance of approximately fifty yards. The Bigfoot are following me like they usually do making enough noise to let me know I am being trailed.

Once I arrive at the location I am disappointed to find that the mock stick structure was undisturbed and my gift of tobacco untouched. I take out my iPod and speakers and begin to play the Native American music and chants. I set up my HD cameras watching opposite directions and then I sit down and eat a snack and take a break.

While I was relaxing the Bigfoot began playing their games again. I heard a small limb snap approximately fifty yards directly in front of me in very thick vegetation. I pick up my video camera and start filming. After about ten minutes I hear a limb snap to my right so I turn and start filming this area very slowly panning left to right. While I am filming this area I get the distinct feeling I am being watched. It is a very strong feeling. Almost bordering on the feeling I get when I am hit with infrasound, but not as overwhelming.

I concentrate on this area for about ten minutes then sit back down. I think I took a couple bites from a granola bar when the feeling something was watching me to my right was overwhelming. I felt almost drawn to this area. I again pick up my camera and start filming the area slowly zooming in and taking videos of the thick underbrush. I am straining to see any part of a Bigfoot. The small size of the viewing screen and glare make it hard to see what I am filming but I keep filming because I know I can review it more closely when I get home. Several times I would film in different directions but I just kept getting drawn to this one area down the hill in very thick underbrush.

I know I am being watched I just cannot find it. I would zoom in and search the woods then zoom back out and search another area. Though I did not see it when I was on location my review of the video once I returned home confirmed my suspicions. There was a Bigfoot hiding in the thick vegetation watching me.

The color algorithm of the digital camera distorted the color, but details of the Bigfoot's face are visible to include the deep set eyes, nose, cheek, forehead, and especially the mouth. During the thirty frames the mouth can be seen opening and closing. There are also some frames where you can see the mid-section or arm through the very heavy vegetation.

The still capture below is the first frame of the thirty frames that I was able to capture.

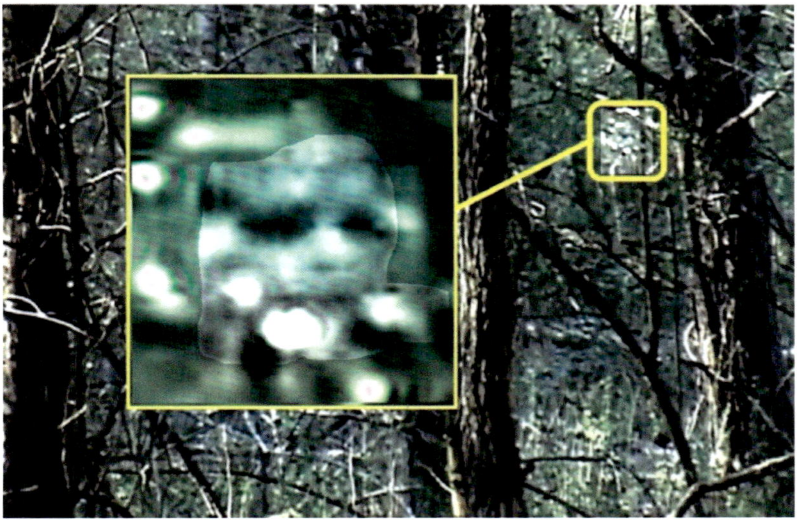

Note in the below still captures the mouth and head movement

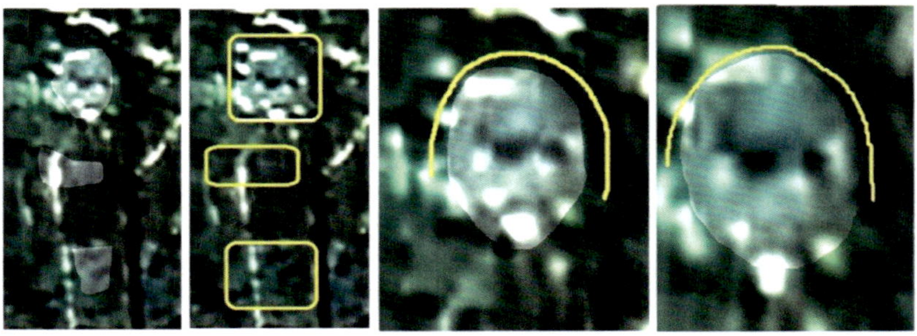

I was able to capture approximately thirty frames of the Bigfoot. Below are select frames in chronological order upper left to the bottom. Take note of the mouth how it opens and closes, is it chewing something? Is it creating infrasound? I noticed the mouth moving and the frames on the next page demonstrate this mouth movement.

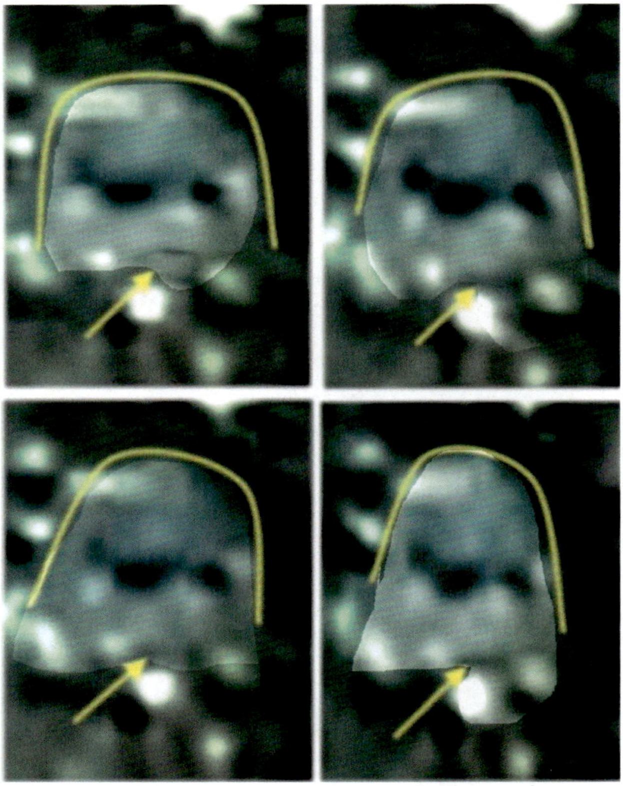

Below is a sketch based on the still captures

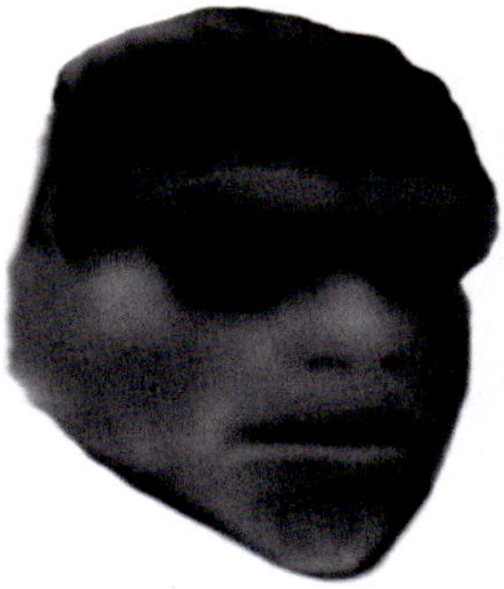

I kept getting the feeling I was being watched from this area. The feeling was so strong that I kept taking video and went into this area to look around. I did not see the Bigfoot while investigating the area. I do not think it was still there, I think it left before I could get into the area.

I finished up and headed home for the day.

Chapter 23 - Tobacco, Bacon, and Bigfoot Watching

Date: February 27th, 2010
Location: Research Area A

I was able to get back in the research area just a week after my previous outing. I arrived in my truck and parked in my usual spot. I was wearing my "uniform" that consisted of drab green pants, camouflaged shirt, and a floppy green "booney" hat. I dressed the same during each outing so the Bigfoot would get used to me and be able to identify me easily.

While I was standing beside the truck getting my gear ready I felt I was being watched. This familiar feeling came over me and I stopped what I was doing and stared into the woods for several minutes trying to see what was watching me. Unable to see anything, I went back to my preparation to hit the trail. I had the Bushnell trail camera with me. It had malfunctioned the last time I deployed it in the research area. I took it home to trouble shoot it and discovered that the issue was dead batteries. I felt dumb for letting it happen. The specification on the camera claim the batteries "could" last up to six months. It had been very cold for several weeks and the cold weather caused the batteries to deplete quickly. So with fresh batteries and a formatted SD card the camera was ready to be deployed. I also brought a plastic bag full of fresh bacon and some natural tobacco to leave as gifts for the Bigfoot.

I walked in the woods and immediately a Bigfoot began following me. I could hear it moving as I moved. Occasionally it would take a few steps after I would stop walking. I must assume it did this to let me know it was following me. After two or three bipedal steps the sound would stop. The Bigfoot would then wait and watch. When I started walking again it would began following again. The Bigfoot's pace always matched my pace. They seem to prefer that I move at a slow steady pace. If I walked at a fast pace it would agitate them. Many of the horse trails go through very thick brush and briar patches. The Bigfoot would always be very loud trying to keep pace crashing through the heavy underbrush as I easily walked along the horse trail with no impediments. This tactic on my part would sometimes invoke a rock throwing response. The rocks are small approximately the size of golf balls and usually thrown in front of me on the trail as if to say "Hey! Slow down so I can catch up!" I also took a little pleasure knowing I could cause them some difficulty since they make such a fool of me most of the time.

With my escort in tow I headed to the Stick Shelter area to check the rocks and the trail cams. When I arrived I noticed nothing had been touched. I took my back pack off and began playing Native American music again. I then proceeded to down load the trail cams while I investigate the stick shelter area. No change in the smiley face or the rocks.

I download the trail cams and then put my Bushnell trail camera back in brush facing the inside of the stick structure.

The entire time I am at the Stick Shelter the feeling I am being watched is very strong and at times literally overwhelming. When the Bigfoot get close to me I can literally feel their presence. I cannot explain phenomenon. I only know in my "gut" I know they are close and watching me.

I moved down the horse trail working my way over to the mock stick structure. It was not long until I had my escort in tow again. I was taking my time and stopping to video any area that I felt a Bigfoot might be hiding. I am on a game trail approximately thirty feet below the old road bed that runs near the area where I have the mock stick structure. I get a few yards away from entering the road bed when I am overwhelmed by infrasound. I stop and begin panning the area in front of me with my video camera. I have a mono-pod so I go very slowly and zoom in and out looking for any sign of a Bigfoot.

I zoom the camera half way into the brush and pause for about two minutes. Then I zoom all the way in to where I think the Bigfoot is located. Unfortunately the area where I zoomed the camera did not contain the Bigfoot. The Bigfoot was just a few feet to the left and up the hill approximately thirty feet. I did not discover this until I returned home and was reviewing the video. I was only a few feet away from capturing close up video of a Bigfoot.

I did capture the Bigfoot but the algorithm of the digital video camera turned the face green to match the surrounding vegetation. I have demonstrated this affect in previous chapters. The deep set eyes are visible and some facial features. The Bigfoot also moves its head to the left then back to the front during the video clip.

Below is a still capture of the area where the Bigfoot was located

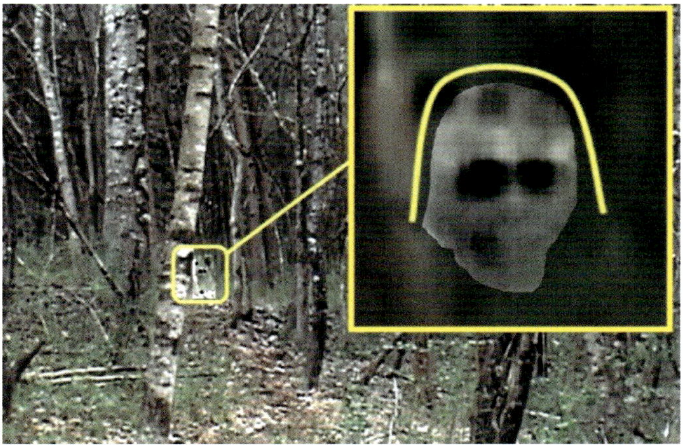

The below photograph shows where the Bigfoot was hiding and the insert photograph shows the area close up. Note the deep set eyes of the Bigfoot.

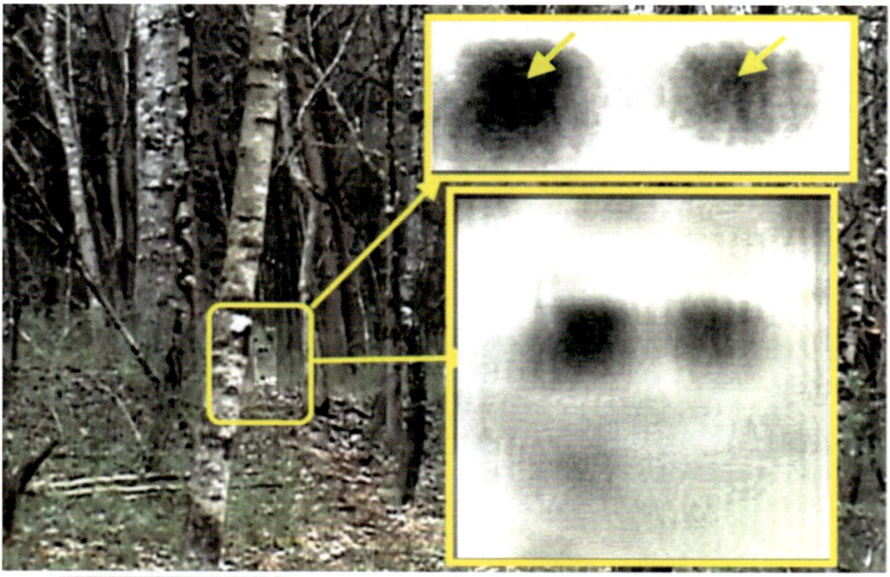

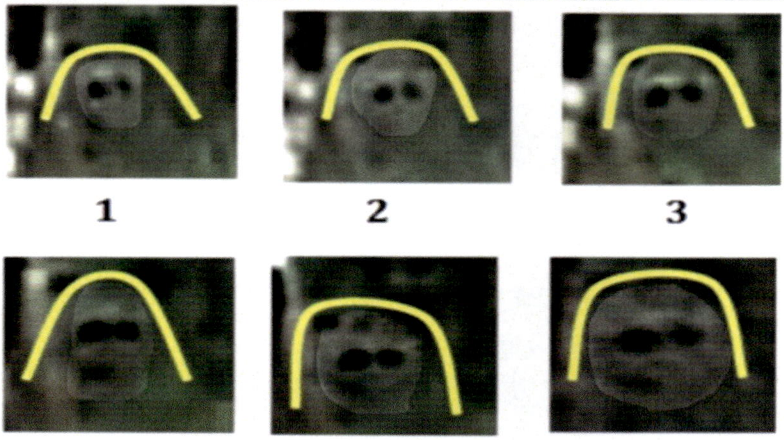

1 2 3

4 5 6

The Bigfoot was slowly leaning out from behind the tree to look at me. The head movement from the start to the end of the video is demonstrated in the above collage.

I concentrated on this area for about five minutes but was unable to see anything with the naked eye. The glare on the video camera view screen made it impossible for me to see the Bigfoot. I moved up the trail approximately twenty yards where the trail began to climb an incline up to the old road bed when I noticed a partial footprint. I could see where the toes had slide in the mud.

The track was still fresh and it showed three toes. The three toes were approximately 3 and 3 ½ inches wide. I compared them with my boot toe and my pocket knife which is 3 ½ inched long.

After I document the toe slide I move up onto the old road bed and unknowingly make my way to where the Bigfoot was sitting. Unfortunately it has slipped away while I was busy documenting the toe slide. Remember I was unaware that there was a Bigfoot hiding in this location. After discovering the Bigfoot when I reviewed the video I went back on a later date and found the location where the Bigfoot was hiding.

Unaware where the Bigfoot was located I move in down the old road bed slowly, filming in the thick areas of brush and vegetation. I cannot always see the Bigfoot when they are in thick cover. They will not move and in the small view finder of the video camera it is difficult to make out any details. Most times I just have to zoom in and pan slowly then review the video when I return home. This day would be no different. I spent several minutes checking the thick brush and I did not see anything while I was in the field. When I returned home and reviewed the video I found a Bigfoot looking at me from some very thick underbrush.

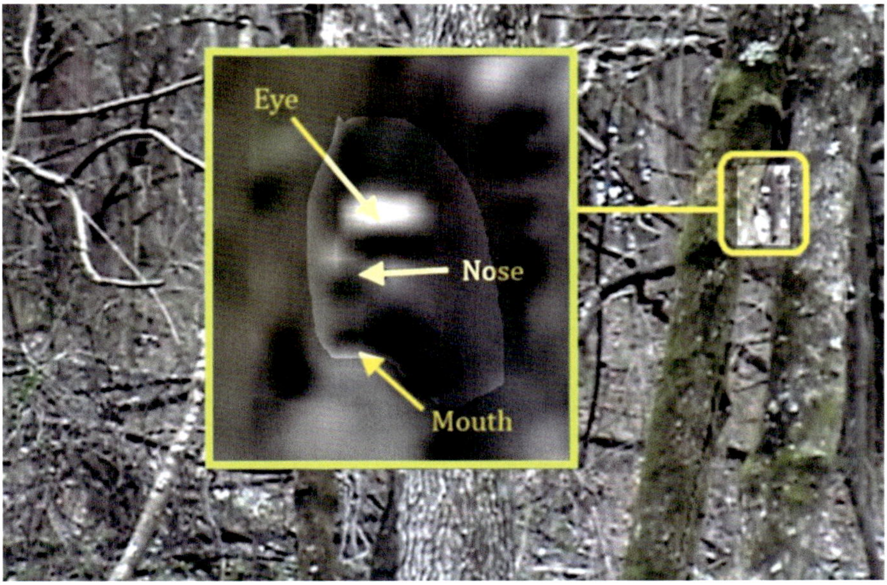

The above still capture shows the Bigfoot peeking at me from thick underbrush

I create a reasonable facsimile of what this Bigfoot could look like

I take my time filming slowly and doing my best to see the Bigfoot my gut is telling me is watching. I finally give up and move up the hill toward my mock stick structure area. I arrive at the area and find nothing has been touched. I set up my iPod and start playing Native American music then I switched out the commercial chewing tobacco with a more natural "twist" tobacco with little or no flavoring (molasses). I also brought some fresh bacon for the Bigfoot. I placed it in a plastic bag with slits and placed it ten feet off the ground. I enjoyed a snack then loaded up, headed back to my truck, and went back home for the day.

Chapter 24 - The National Park Bigfoot Research Project Begins

Date: March 6th, 2010
Location: Research Area B

I am fortunate that the area I live in is close to prime Bigfoot habitat. I found through researching Bigfoot reports that the Research Area B is a "hot bed" for Bigfoot activity. I wanted to branch out and find another good area to research. The park encompasses 500,000 acres of forest land so I needed to find a location that could attract and hold a Bigfoot population. My experience at Research Area A and some research on Bigfoot sightings brought me to the following conclusions regarding a good area for Bigfoot.

- If the area can support Black Bear then it can support Bigfoot.
- The Bigfoot are partial to horses and areas with horse trails.
- The Bigfoot are drawn to women and the sound of playing children.
- A larger percentage of women were experiencing their menstrual cycle when their Bigfoot sighting occurred.
- Water sheds and areas near running water.
- Ridge lines.
- Forested areas adjacent to or near small towns or housing developments.

I started looking at maps and aerial photos of the Park, concentrating on the areas not far from my residence. I found an easy to moderate hiking/horse trail that had heavy hiking and horse riding traffic. The trail was also along a large year round stream. The area had direct access to the small tourist town and to the more remote areas of the park. I would also later find out the many women and families with children frequented the trail. It was literally a "Bigfoot Magnet". This area contained all the characteristics that made it prime Bigfoot habitat. Now all I had to do was get out there and start sniffing around.

The weather had turned cold since my last outing and my family was suffering from a case of "Cabin Fever". The day was cold but the sun was bright and the sky was deep blue so we decided to take a short hike. The area I had found through my research would be perfect for an afternoon hike with my children. The trail was relatively easy and since this would be my first visit to this trail I would be able to poke around and do some scouting for my research expeditions. A plus would be my children. They would be a natural attractant to any Bigfoot in the area. I know the odds of having a Bigfoot sighting are very slim, but having my children along could make the difference.

We loaded up in the truck, grabbed the video camera and off we went into the Park. My wife's descendants are from an area in the park. We visit this area on a regular basis because of its natural beauty. This would be the first time I would enter the park as a Bigfoot researcher but it would not be my last.

We arrived at the trail head around 10:00 AM. The temperature was in the high 30's and had been in the low twenties the night before. The ground had just begun to thaw out. There was still snow on the ground from earlier in the week. My two younger sons had joined me on this hike, my wife decided to read a book and stay near the trail head in the warm car.

The boys and I started up the trail and I noticed how thick the underbrush and laurel thickets were on both sides of the trail. The trail follows a small fast moving creek with high ridges on both sides. It was easy to see that the step ridges had many areas where a Bigfoot could hide and observe the activity on the trail without being noticed. The children are walking ahead of me playing and being boys, throwing rocks in the stream, being loud, and just generally being rambunctious. Then they both let me catch up to them and I begin to get that familiar feeling. Before I can say anything, my twelve year old, looked up the hill side as says "Dad, I am getting the creeps, something is up there". I filmed the area for a couple minutes as we both looked for what was watching us.

In the field we did not notice anything; we just had that gut feeling we were being watched. When I returned home and reviewed the video I found the highlighted anomaly in the above still capture.

I was able to pull out a few good stills of the "Bigfoot". The "face" is not clear but in my opinion it is a Bigfoot. It has the characteristics including deep set eyes and gray skin. The below photograph has an actual still capture on the left and my interpretative drawing on the right.

We looked in this area for several minutes unable to see anything and decided to move on up the trail. I find it very interesting that both my son and I sensed that we were being watched by something. My son asked if there was a Bigfoot up there watching us and I told him "I cannot see it but I know one is up there". "Think it will follow us dad?" my son asked. "Probably, he will shadow us while we are hiking, but I would not worry about it maybe we will get a better look at it if he gets more curious. So just play and enjoy yourself and don't worry about it." That seemed to make my sons feel better and they went back to playing and throwing rocks into the stream.

We worked our way approximately a mile back into the mountains where there was a trail intersection. I decided to take the left fork and as we were going around a mud hole I noticed a footprint. The ground had only been thawed enough to take a footprint for an hour at the most!

This footprint was made minutes before we arrived at this location. The footprint was just of the ball of the foot and toes. The subject that made the footprint was running!

Please note that we were the only people on the trail, no one had passed us nor had we met anyone. Also the ground was frozen from the night before and had just begun to thaw, this track was no more than an hour old. The track was left approximately one hour before we found it. Also the air temperature was approximately thirty five degrees.

This would be the first of three tracks we would find on this trail. After documenting the track we start walking up the trail again. We start up a steep incline as the trail narrows. We round a sharp turn in the trail when I notice another footprint on the hiking trail in a muddy area.

This track is larger than the first smaller track. We now have two individuals walking the trail. In the below photograph I compare the footprint to my boot and my bare foot to show the size difference.

The below photograph shows my foot over the print, my foot is four inches wide, and the track is over six inches wide.

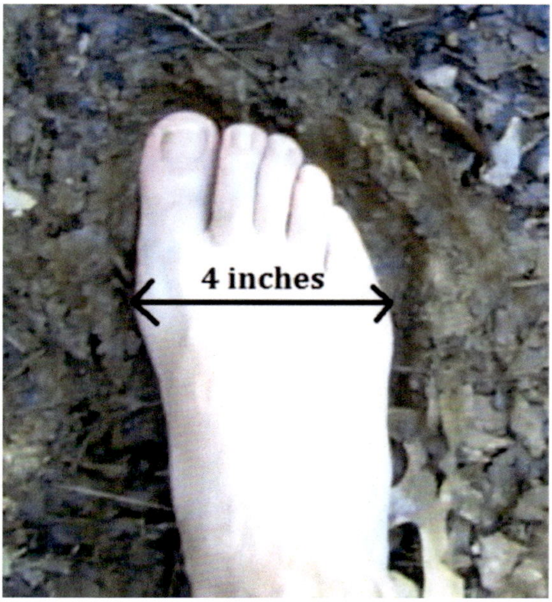

Note how the Bigfoot print is "boxy" and the toes are more uniform in length than mine.

We continue on up the trail and approximately thirty yards from the second footprint we find a third partial footprint. The Bigfoot appears to still be running and only the toes and ball of the foot are visible.

The boys and I continue up the trail after documenting the third foot print. The trail is now working its way along a high narrow ridge and has become very narrow with laurel thickets on both sides of the trail.

We were nearing the top of the ridge and the boys complained they were getting tired. About that time I began getting an odd feeling. It was more than the now usual "I'm being watched" feeling. So I told my boys to stay where they were and rest and I was going up the trail just a little more.

I went up the trail approximately thirty yards but kept the boys in sight when I experienced very intense infrasound. When it first hit me I had trouble getting my breath. I also started getting the message "You need stop leave now". I then began to have trouble focusing my eyes. My eye sight became blurry and my eyes began to water. I immediately turned the camera on my face and started to describe what I was feeling and how I was being affected by the infrasound.

I decided it was time to leave since my children were near and it was obvious the Bigfoot did not want me around. I videoed my back trail as I left. One interesting thing is that a couple months from this outing the Bigfoot will leave a stack of rocks for me with a wild mint plant under the top rock.

I went back down to where I left the boys and we made our way back down the trail. Once we came back down off the mountain I stopped and investigated the natural funnel I had spotted on the way up the trail. This area had a creek running through it and I found game trails that had bear, deer, and wild boar sign on them. I knew this would be an area I would need to investigate further.

After investigating the creek area we made our way back out to the parking lot area and returned home.

Chapter 25 - A Thief, Bigfoot Smells and Missing Bacon

Date: March 20th, 2010
Location: Research Area A

The sun has been up for just a few minutes when I arrive at the research area. I load up my gear, get my camera ready, and start down the trail. The wind is up today and swirling so when I started smelling the "Bigfoot Smell" I was not surprised. The Bigfoot always stay down wind of me but today the wind gives them away.

This was also the first warm sunny weekend of the spring and I could "feel" the Bigfoot were back in the area. The sounds, the smells, the hair raised on the back of my neck, the odd out of place bird call, all of these things add up to let me know the Bigfoot are back.

It took me some time to make it down first hundred yards of the trail because the Bigfoot were making themselves known today. I made it to the dry creek bed and found the below rock pile and sticks laid out in a pattern.

I do not know what it means if anything, I only know that a couple weeks ago these rocks were not on the trail. Did a Bigfoot do this or was this just another "coincidence"?

I keep moving down the trail taking my time and panning the thick brush with my video camera. I was hoping that I might be able to catch a few frames of a Bigfoot spying at me from the thick underbrush. I knew they were very close but were being careful to stay out of sight. I came to a large mud hole in the trail when I noticed the Bigfoot smell. It was literally overwhelming. It was very potent and so strong that the Bigfoot had to be close to my location.
I decide to go check out the area around the "hide" that is located beside the lake. The "smell" seems to be coming from this area, plus I had put some black thread across several trails a couple weeks ago at the seven foot level and I wanted to see if a Bigfoot had walked the trails and broken the thread.

I made sure the thread was at least seven feet off the ground so only a subject that was seven foot in height or greater would break it. If the thread is broken then this will tell me the Bigfoot are back in the area and using the trails to travel again.

I checked the thread placed on the trail in front of the hide and it had been broken! I check the area and find no other sign of activity. I then work my way back out to the main trail. I ease my way to the Stick Shelter area to check the trail cameras and the rocks. When I approach my first camera I can see that it has been messed with. It is just hanging loose by the security cable. Did a Bigfoot do this or was it something else?

I turned around and looked into the shelter and noticed my other camera was also hanging from the security cable. Could this be what I have been waiting for? When I down load the SD cards will I see one or more Bigfoot messing with the trail cameras? I notice that the entire shelter area has been disturbed, rocks have been moved around, sticks broken, and the ground disturbed. I booted up my laptop and started the download anticipating video of a Bigfoot.

When the video is downloaded and I review it and instead of a Bigfoot I find an elderly man, who came in by boat, attempting to steal my cameras! He went to the Bushnell first, when he could not get it loose he went for the camera in the shelter. He completely contaminated the research area.

I decided to move my cameras since the area had been compromised. I would leave one camera at this location, place one on a nearby trail, and move the Bushnell to the mock stick structure location. I leave the stick shelter and head toward the island to check an area where I have put thread over the trail. I have an ongoing theory that the Bigfoot use this little island as a bedding area. I have placed thread over the trail that leads from the mainland to the island.

When I check the thread I find it has been broken. It was placed seven feet off the ground. I think this confirms that the Bigfoot have moved back into the area and using the island.

I then work my way over to the mock stick structure location where I had left fresh bacon in a plastic bag. I find the limb that I used to suspend the bacon has been pulled down to the ground. The end of the limb is broken and the plastic bag empty

The bag was completely empty and ripped. The interesting thing about the bag was its condition. The top was unzipped and open. Did a wild animal climb up the tree, walk out on a limb that would not support its weight and eat the bacon? Did a Bigfoot reach up and grab the bag, open it up and eat the bacon? Was the plastic bag discarded on the ground where it was found by a small animal, ripped opened and licked clean of the bacon grease?

Another interesting item is that the plastic bag was pulled off the end of the limb not ripped off the limb. Whatever removed the bag slid it off the end of the limb. Would Raccoon or Possum do this? Would they not just rip the bag off the limb? I find more questions than answers. Did a Bigfoot reach up, slide it off the end, open it, and then consume the bacon?

I decided to hang a trail camera in this location to hopefully catch a Bigfoot moving through this area. I placed it on the side of a cedar tree watching the area where I had put the bacon. I did not place it watching the mock stick structure. I was hoping to decoy the Bigfoot with the camera and trick them into taking my gifts on the mock stick structure. The Bigfoot seem to pride themselves in out smarting me so I thought I would use this to my advantage. Of course out smarting me and staying hidden is part of their nature.
I finish up at the mock stick structure location and then make my way to the next trail camera I have setup. I ease down the trail and notice an odd swatch of hair lying on the side of the trail. The hair is black and gray and it appears to have been cut with a dull instrument. The edges of the cut hair are not uniform and appear "jagged". Many Bigfoot sighting witnesses have reported that the bangs of the Bigfoot appeared uneven like they had been cut with a dull instrument in order to get the hair out of the eyes. I was unable to test the hair for DNA since there were no roots or skin tags.

I bagged the hair and made my way to the final trail camera. I checked it and there were only videos of horse riders passing by on the trail about 30 feet away. This cedar tree would be a perfect location to watch the trail and stay concealed.

I checked this camera and left the research to return home.

Chapter 26 - Mock Charges and a Bigfoot watching me - Again

Date: April 10th, 2010
Location: Research Area A

The day started off very early as I made it to the research area just after the sun had risen. The day is bright and warm. Bird songs fill the air as I put my gear on and get ready to hit the trail. I pull on my back pack, get my video camera and hit the woods. I do not make it very far into the woods before I began hearing movement to my left rear; as usual it is the sound of bipedal walking down wind of my position. The Bigfoot is not close enough to see but makes just enough noise to let me know it is there and following me. I stop several times and slowly pan the camera 360 degrees hoping during review I will have captured the Bigfoot on video.

I strain to make out anything that looks out of place. I look for an eye, nose, cheek, arm, hand, part of a leg, part of a chest, anything that would give the presence of my escort away. You must train your eyes to look for body pieces. The Bigfoot will never expose themselves out in the open unless the circumstances are dire for them. I therefore look for parts of the body that may be exposed while I am being watched. In my experience the Bigfoot will hide its entire body then slowly move its head out through the vegetation to take a peek at me. They also will camouflage their faces with leaves or other vegetation. They seldom expose their entire face. They will peek out from the back side of a tree or through vegetation with just one eye. The eyes are so deep set that they are black and recessed. I have even observed a Bigfoot closing its eyes in order not to make eye contact or show the whites of its pupils. They know eye contact means recognition.

I made my way to the stick shelter area with my escort in tow. I arrived at the stick shelter and looked around. There did not seem to be much activity with the rocks or smiley face. I started my trail camera down loads and set up two video cameras to watch the brush around me as I investigated the area.

While the cameras are down loading I look around area for sign or any activity. I notice something interesting that I had missed before. When I looked near the thick brush on the north side of the stick shelter area I find what looks like a "star" stick structure.

It can be difficult to determine if a stick structure is legitimate or naturally formed. I put this one in the 50/50 category. I also find sign of heavy footfalls with broken limbs on the ground. A Bigfoot is very stealthy but they cannot float through the woods. You can find evidence of their passing through the woods if you know what to look for. They do break limbs and leave depressions on the ground with their heavy body weights. They will bruise vegetation or compact the matter on top of the ground. The problem is they know where to walk and how to walk to minimize leaving behind this sign.

The result of this practice is sporadic instances of their footfalls. It makes it almost impossible to track them with any consistency.

I leave the stick shelter and head toward the mock stick structure area. I do not hear anything following me which is odd given the recent events but I continue on my way filming and walking at a leisurely pace. When I get to the mock stick structure it appears my tobacco gift again has been left untouched. I move over to a big smiley face I made on the ground to see if anything has happen to it. I do notice a couple rocks in the formation have been moved around. It always amazes me how I can place an isolated group of rocks in the middle of the woods in my research area and they will be moved or disturbed. I have actually gone out in the woods in a random location not in the research area made a rock pile. Each time I have returned to these rock piles they have remained untouched. In my research area my rock formations get moved around all the time.

After setting my cameras out to watch the brush I get to the business of down loading the images from the trail camera. I also sat down to eat some lunch and rest. While I was sitting down eating I hear a loud crash to my right. The sound seemed to be coming in my direction and whatever was making the noise was very large. I grabbed my camera and prepared myself to capture whatever was coming through the woods. Moments before the unseen subject was within visual range it just stopped and the woods went completely silent. I stood up and started filming the area where I heard the noise. Approximately five minutes later I hear bipedal walking close to me. I concentrate my filming in this area. Approximately five minutes later I hear bipedal walking again and I began to get the distinct feeling that I was being watched. I could literally “feel” their eyes on me. I zeroed in on an area where I “felt” they were located. I strained to see if I could locate the Bigfoot. I knew they were close. I filmed, then moved closer, filmed more and moved even closer. I then began to smell the “Bigfoot smell”. I knew they were close but I could not get them to move or show themselves.

Upon careful review of the video I noticed one subject. Below is the still capture with enhanced inserts.

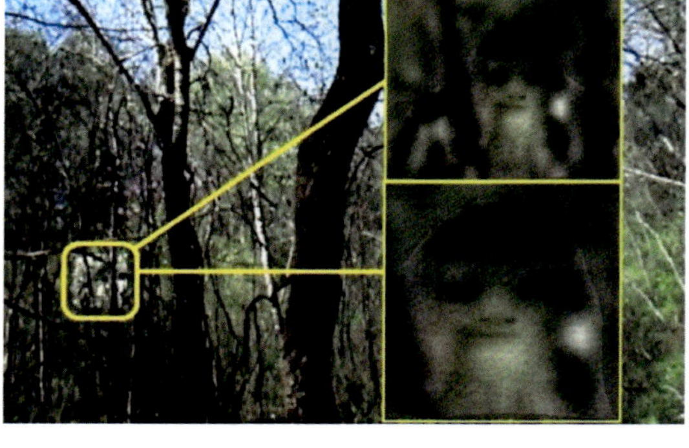

I think you can clearly see the deep set eyes and the nose. This still capture is from the area where I heard the sounds and felt that I was being watched from.

I moved closer to this area and kept filming. I did not see the following face while I was on location because of the sun's glare on the view screen but it was not difficult for me to identify when I reviewed the video.

The face is in the below still capture. This was located in the exact same area as the previous photographs only now I am closer and the sunlight is reflecting off the face. In my opinion the Bigfoot choose to go down wind of me, but this put them with the sun in their faces which worked to my advantage.

The following montage shows the few seconds I captured the "face" from frame to frame. The Bigfoot is rocking back and forth. I had the camera mounted on a mono-pod so the subject was doing the moving in the photographs. Note how the left eye will go behind the tree then be completely visible a few frames later.

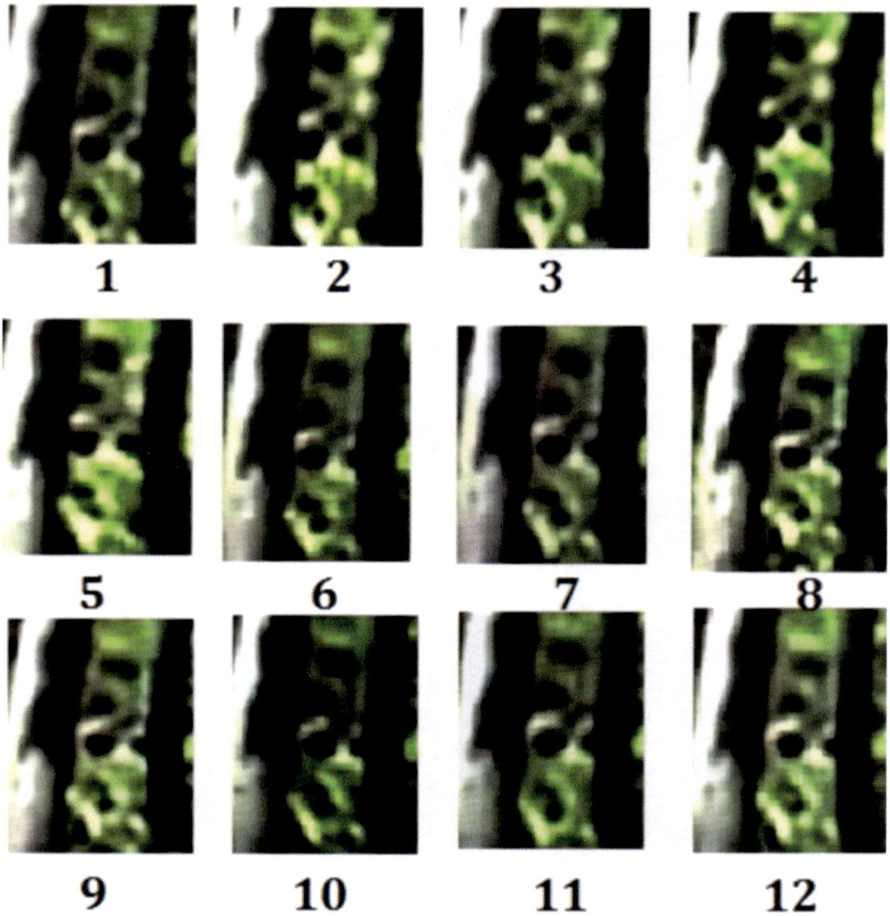

After I filmed the area for at least fifteen minutes I move back to the mock stick structure and start packing up my gear. I reset the trail camera and made my way back toward my truck to leave the area.

Chapter 27 - Sneering Bigfoot and New Camera

Date: April 30th, 2010
Location: Research Area A

After several months of saving money I was able to purchase a new high definition video camera. Up to this time I had a good Sony Handycam Nightshots camera with a 30GB hard drive and two inexpensive off brand high definition cameras. I selected the Sony Handycam with an 80GB internal hard drive. This camera will allow me to be in the field a full day and never shut my camera off or have to use one of the lower quality cameras. The HD will give me very clear and crisp video.

The old camera served me well but it was not high definition. The new camera should improve my ability to document my findings with clear crisp video and photographs.

I made this trip late in the afternoon instead of early morning. I thought I would change the time of my outing to see if there was any activity late in the day. I arrived, put on my back pack, prepared my new HD camera and hit the trail. I made my way down the trail experimenting with my camera and investigating the new features. The zoom feature is very powerful, the picture quality excellent, and with the auto-stabilization program I knew I made a great choice!

While enjoying my new camera I almost trip over a dear carcass stripped to the bone by scavengers. I wonder if this was originally a Bigfoot kill for the nutrition rich liver and kidneys. I have not found a fresh Bigfoot kill but I have seen several pictures and read numerous accounts of how they dispatch a deer and remove the organs to consume.

I documented the deer carcass then made my way down the trail toward the stick shelter. The area was relatively undisturbed except for the smiley face. It had rocks missing again making a Y pattern. The interesting thing about the smiley face is that I have placed it where only slight weight or pressure would overturn it. I also used the rocks to balance it, so moving the rocks around would overturn it. During the time that I have been using the smiley face it has never been overturned. If a small animal like a raccoon, possum, or squirrel was moving the rocks around at least once they would have overturned the smiley face, but it has never happened. Would a small animal move the rocks around but make sure they balance so not to overturn the smiley face?

This pattern of a Y is unique and the first time since I began researching this area I have seen this made with the rocks. This could be another way the Bigfoot express the 3 -1 pattern. In the next sequence of photographs I will show how the rocks were moved from a smiley face to the Y pattern.

After I downloaded the trail camera at the stick shelter I checked the trail camera I placed to watch the back trail into the stick shelter area and found something very interesting. The trail camera's lens is blocked by an insect cocoon! This is very odd. I have had multiple cameras out for many months and nothing like this has ever happened.

The nagging question is did a Bigfoot do this or did this insect select this location? It is just another one of those coincidences that happens in this area that make you wonder.

After I download and reset the camera I go back to the stick shelter area, gather my equipment and move back out to the trail to check for any activity near the island. I am still learning how my new camera's auto focus functions as I approach the island and begin slowly panning the opposite shore.

I zoomed in and out of the marked area several times before I inexplicably turn and walk off. When I returned home and reviewed the video I found there was a Bigfoot hiding on the island.

It was peeking out from around a tree. The most disturbing aspect of this encounter is I have no recollection of taking this video or seeing the Bigfoot. Only after review of the video did I see the Bigfoot sneering at me from behind the tree. I think exposure to infrasound produced by the Bigfoot is the explanation for why I did not react to seeing the Bigfoot. I also think it is responsible for my memory loss

during the encounter. I was only able to capture approximately thirty frames of the Bigfoot but you can clearly see the exposed teeth. The mouth then closes at the end of the sequence.

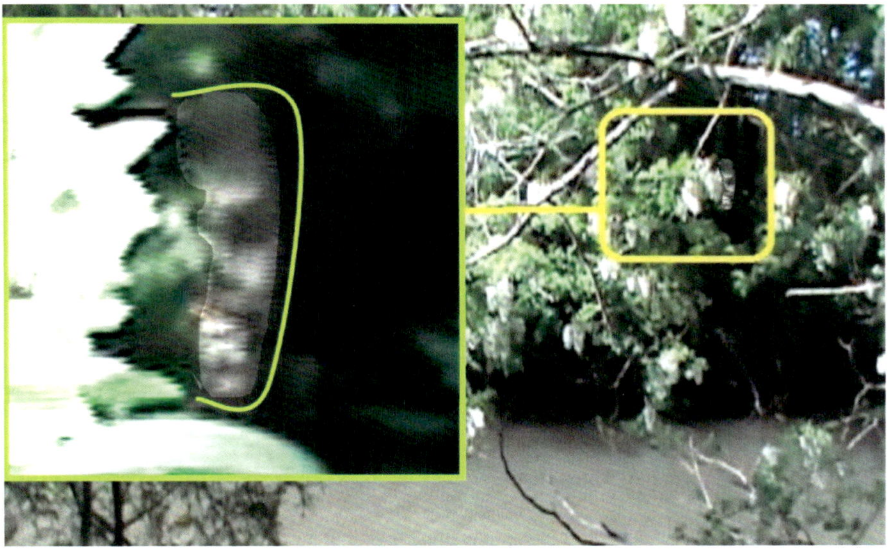

The above composite photograph shows the general area and the insert shows the Bigfoot with its teeth showing

When I zoomed all the way in the auto focus blurred everything so I had to pull back out and then zoom in about fifty percent. I do not remember taking this video sequence. I must have been aware something was there or I would not have zoomed in and out more than once. The mystery is I do not remember taking the video because of the influence of what I believe was infrasound.

When I zoom back in the second time the Bigfoot is now in focus and I am able to catch the end of the sneer with the mouth closing.

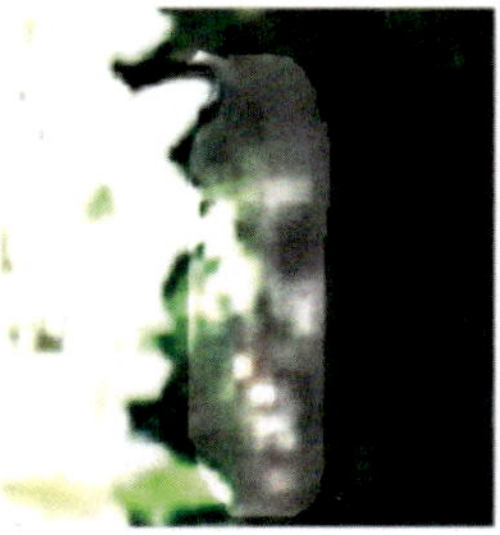

1

2

3

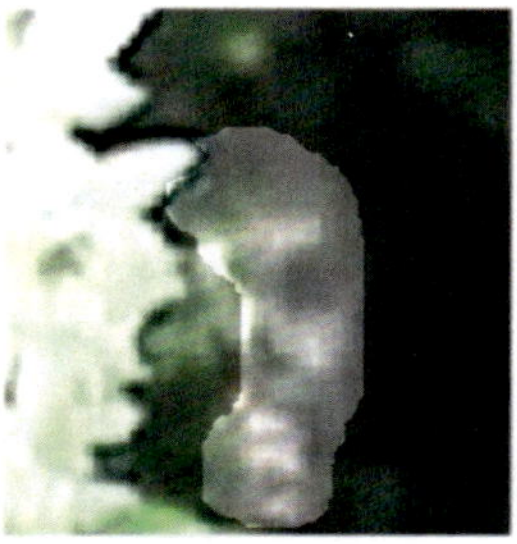

4

5

6

7

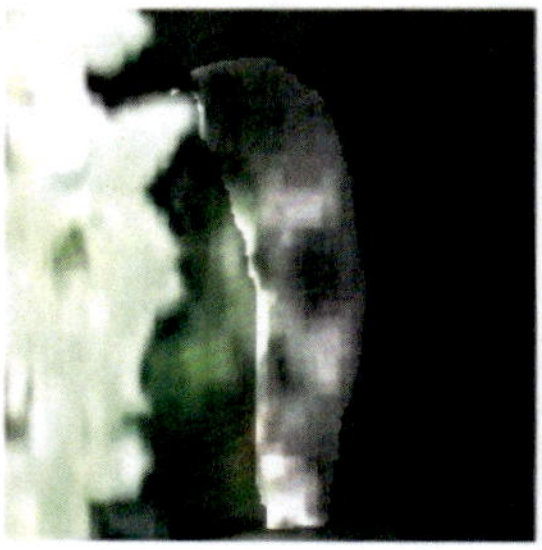

8

9

10

During the previous sequence the Bigfoot was closing its mouth. The two photographs below show the first and the last picture in the sequence to demonstrate the mouth closing.

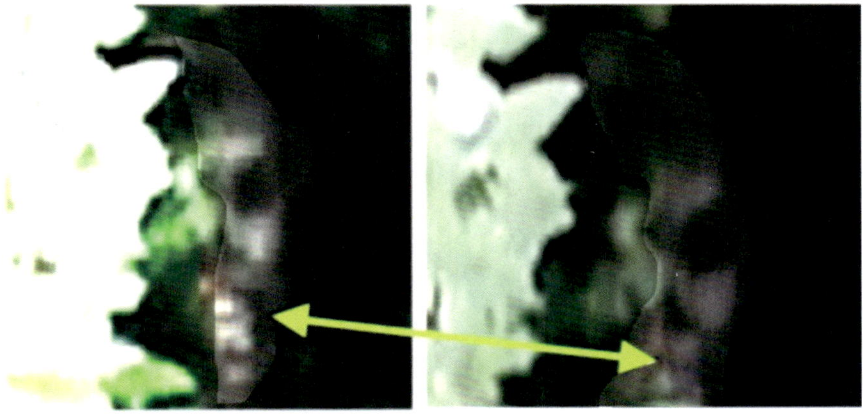

1 10

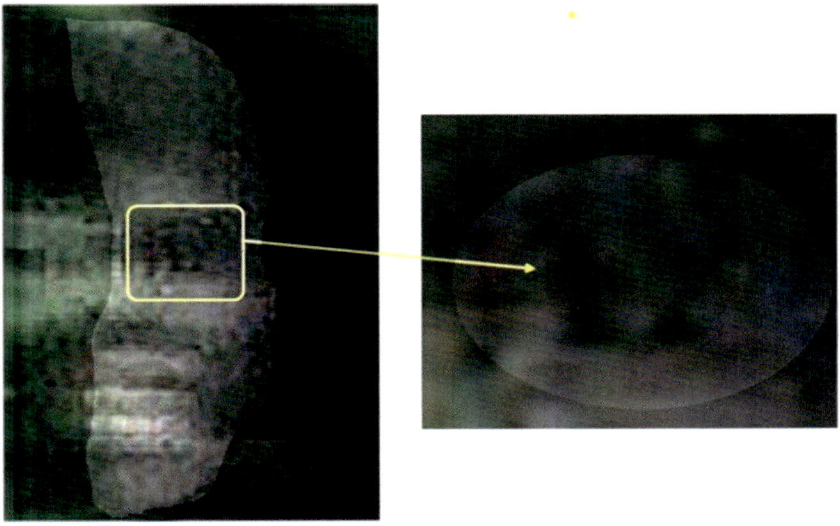

In Frame 10 the eye of the Bigfoot is also visible. In the above close up you can see the detail of the left eye.

The following sketches are based on the actual photographs. Each sketch shows a different position of the mouth, from a full sneer to a closed mouth.

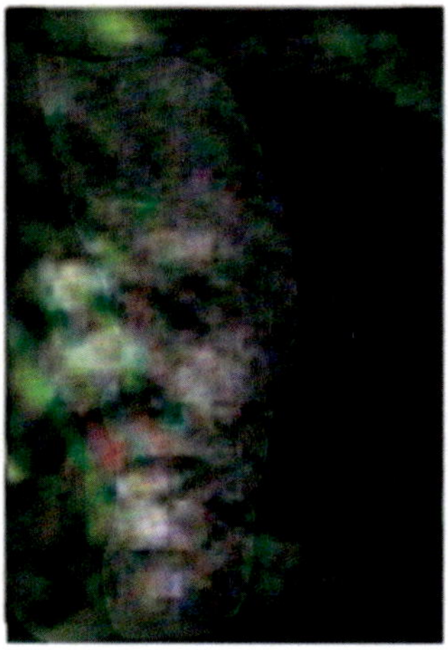

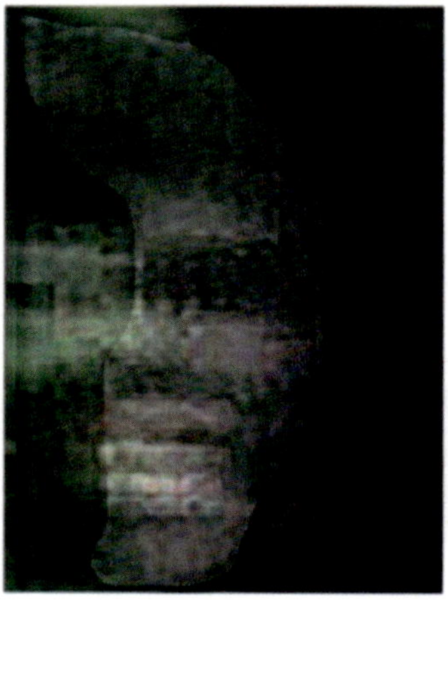

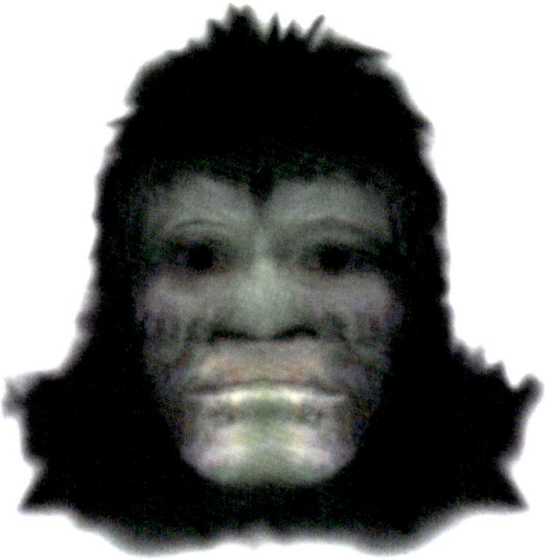

I think the previous composite enhancements are a reasonable facsimile of what the Bigfoot actually looked like in full face view. I get frustrated not being able to get them in full view, but the nature of the Bigfoot make it extremely difficult. They are always in stealth mode, hiding and peeking. They never expose more than one eye when they peek. If someone sees them full body, close up and face to face it is rare.

I was going through the sneering Bigfoot video clip frame by frame when I noticed for a few frames you can see the Bigfoot's left hand moving.

In the above composite photograph what appears to be the left hand is visible

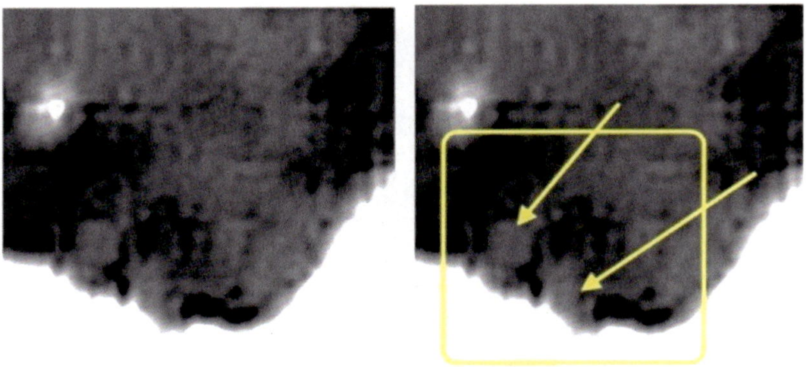

The fingers nails are visible in the above still capture

During the analysis I could also see the fingers moving. The below composite demonstrates the finger movement.

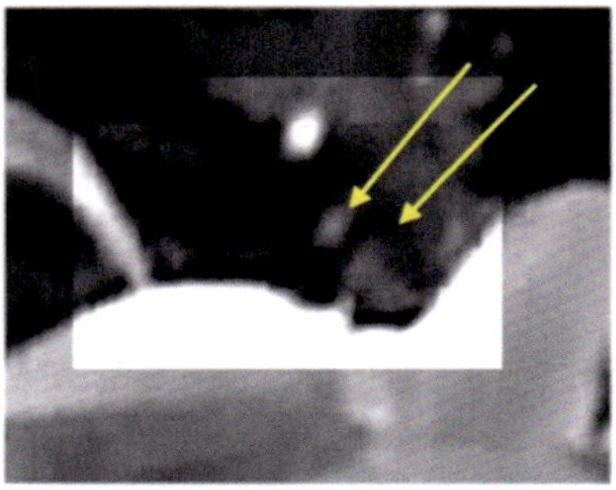

Frame 1

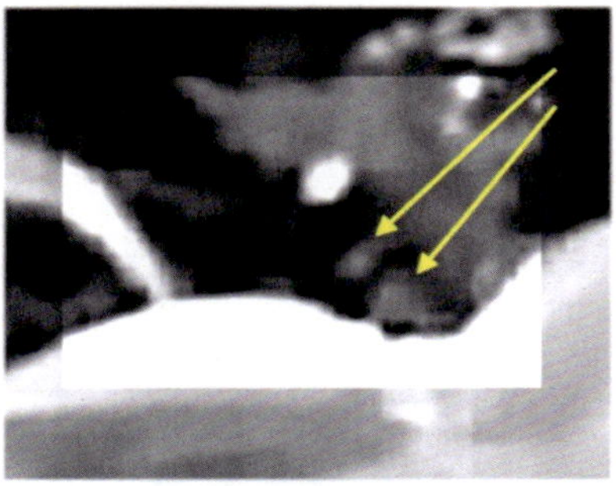

Frame 2

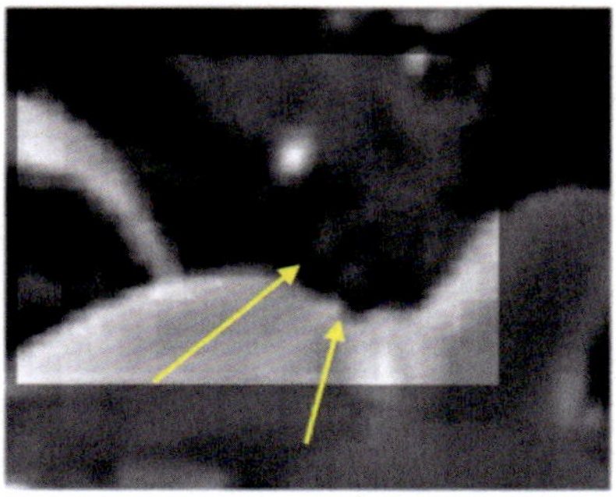

Frame 3

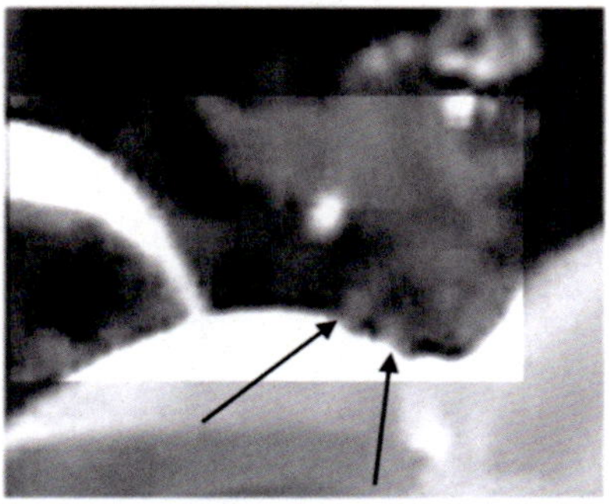

Frame 4

I went back to the area after discovering the footage to do detailed follow-up. By doing measurements I was able to determine that the Bigfoot was approximately seven feet tall. I placed orange marker tape on a sapling that was beside where the Bigfoot was standing to determine the height.

.

The above composite shows the seven foot mark hits the Bigfoot on the top of his forehead

While I was on the island investigating the sighting location I notice what appeared to be "finger nail" scrapes in a limb that would have been directly in front of the Bigfoot. These "finger nail" scrapes line up with the area where the Bigfoot's hand and fingers were visible.

The above still capture shows a long scrape on a limb that was in front of the Bigfoot

The above still capture shows what appears to be where two fingernails dug into the limb

After I wander off unaware I have captured the footage of the Bigfoot I make my way toward the "Mock stick" structure location. I am hopeful that my tobacco has been taken and that my trail camera in that area may have captured something.

When I arrive I find that the tobacco is still untouched. I am wondering if this group of Bigfoot knows they are supposed to like tobacco. I say that in jest because I am finding that Bigfoot are definitely unique individuals. They may have some common behaviors but just like everyone else they have likes and dislikes.

I also check the rock smiley face and it has been disturbed. I check it closer and notice hoof prints in the soil. It appears that horse riders have found their way off the trail and back into this area.

When I downloaded and checked my trail camera my suspicions were confirmed. A pair of horse riders had wondered off the main trail and found my research area around the mock stick structure. The horses trampled on the smiley face.

After I finished with the download of the trail camera I go to my last camera that has been setup to watch the area where I captured the Bigfoot hiding in the Cedar tree and brush near the horse trail. The foliage was becoming very dense and now I could understand what an excellent hiding place this area made. The Bigfoot could watch the trail from this area and remain completely concealed.

After I checked the trail camera I started heading out of the area. It was getting late in the day and I wanted to get out of the area before dark. I noticed what I call a Y marker in a limb just off the smaller trail as I was making my way out to the main trail. This is the third time I have found this type of "marker". I think that the Bigfoot are finding a limb that is six to ten feet off the ground that forks on the end to form a Y. They break this limb off then hang the broken limb on the broken remnant or on another limb close by forming a marker.

After documenting the Y marker I made my way out to my vehicle to leave the area.

Chapter 28 - Exploring a New Area While Bigfoot Sneaks Around Me

Date: May 1st, 2010 and May 8th, 2010
Location: Research Location B

I decide to take some time and explore new research areas. The National Park covers over 500,000 acres in two states. I have plenty of "new ground" to explore. On this day my family and I go the National Park to enjoy the warm weather and outdoors. During our two trips I took my camera and was able to sneak away for some hikes along a major stream in the area. I thought that since Bigfoot are attracted to playing children and women this would be the ideal location to look for a Bigfoot. The area and the trails were inundated with families.

We started our day out with a lunch at a beautiful picnic area in the National Park after lunch we decided to drive into another tourist area and go around the "loop" road that meanders through this area. On the way to our destination we encountered what locals call a "Bear Jam". A Black Bear has wandered close to the road and all the tourist are stopping their cars and taking pictures. I could not resist taking some video as we slowly drive by the Bear feeding beside the road.

After we made it through the "Bear Jam" we went into a popular tourist area that has a road that encircles it. This location is prime Bigfoot habitat. There is abundance of wildlife, water, horse trails, women, and playing children. They have direct access to vast wilderness area and a lush green valley to live in.

We stopped at a pull off and I decided to go for a short walk with my camera up the hill. I found the woods abundant with wildlife on this warm spring day. I was not fifty yards into the woods when a flock of wild turkey walked across the trail in front of me.

After a short walk we got back into the truck and made our way around the loop road. When we came to the half-way point of the loop road, we took some time to stretch our legs, and just sit on the ground and relax for a while. I decided to "sneak" off and take a walk down a trail along the fast flowing stream. The trail is regularly traveled by visitors of the park and well worn. Across the stream the foliage was thicker and not many people ventured across to this side.

I walk down the trail for approximately thirty minutes enjoying my walk but still filming. A couple times during the walk I got the familiar feeling I was being watched. I would stop for a few minutes and search the woods and use my camera to zoom in and look for any sign of a Bigfoot. This must have aroused their curiosity. Once I reached the point where I decided to turn around and head back I was being followed. I started pointing the camera to my left filming the area across the creek as I continued to move forward. When I reviewed the video at home I found that I had a Bigfoot paralleling me through the foliage. In the time lapse composite photograph below there are six frames in sequence of the Bigfoot following me.

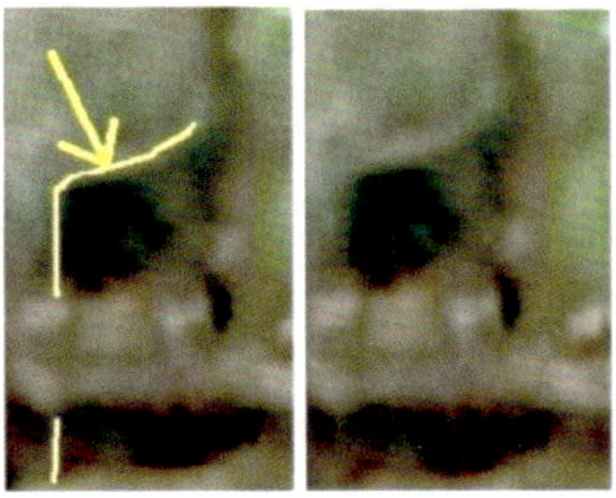

Frame 1

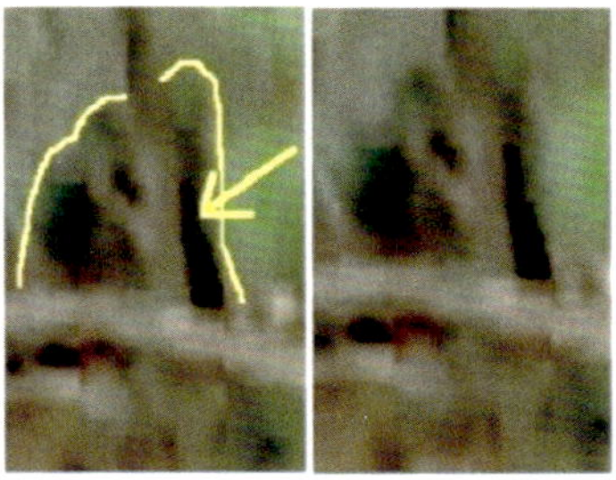

Frame 2

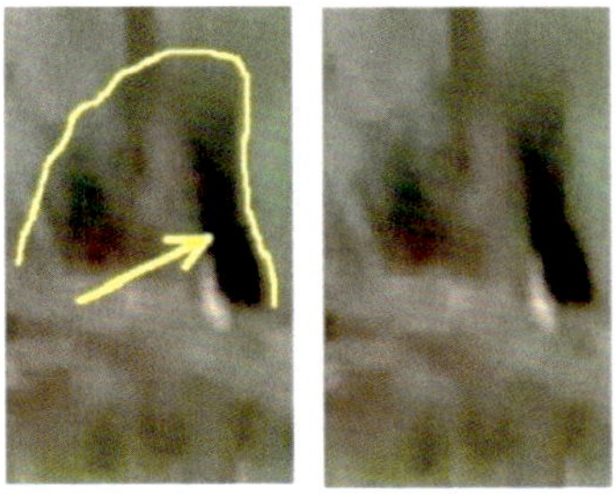

Frame 3

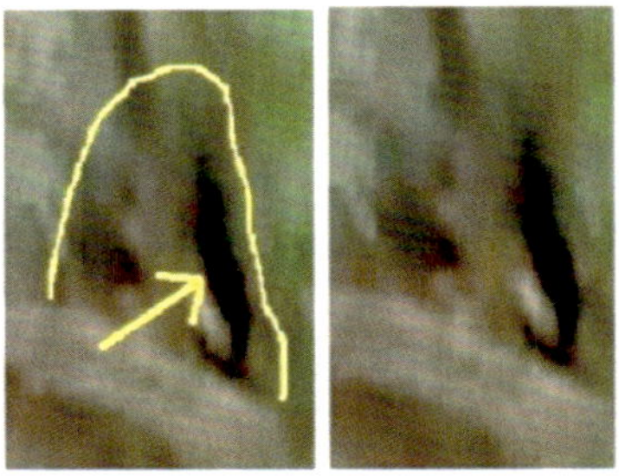

Frame 4

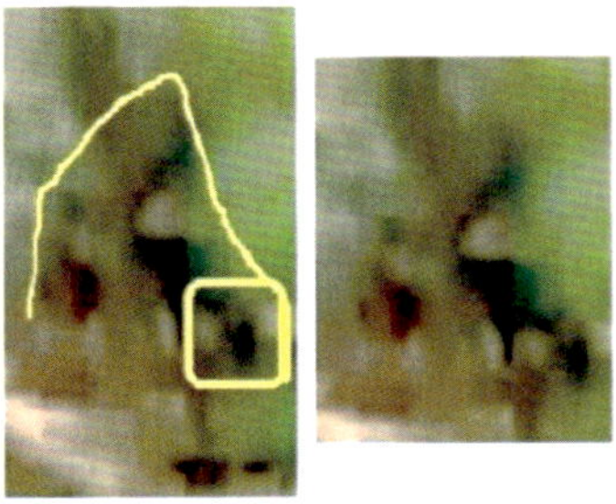

Frame 5

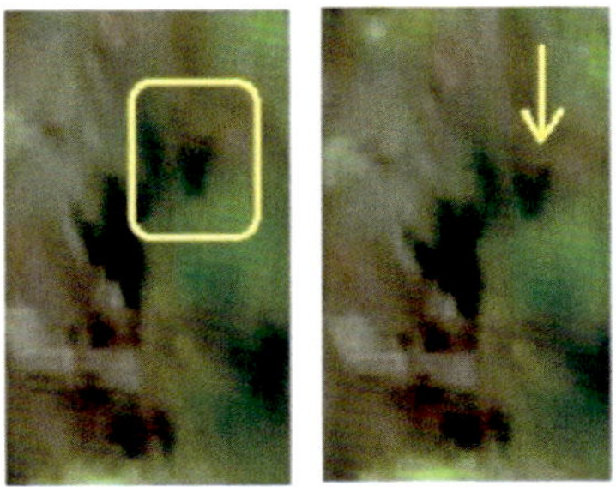

Frame 6

The Bigfoot is stepping over a large log to keep up with me. The speed at which this Bigfoot moves is amazing. The video camera records at thirty frames per second and in just six frames it moves up and over a down tree into the thick vegetation. Frame 1 is the Bigfoot bent over. In Frame 2 the Bigfoot is now upright but hunched over slightly and you can see its massive right arm hanging down. Frame 3 demonstrates the slight swing of the right arm forward as you can see light between the arm and the tree. The Bigfoot is also bending over slightly more than the previous frame. Frame 4 shows the arm moving forward and now you can see the right hand and the right arm. Frame 5 shows the Bigfoot putting his hand on the large log that is parallel to the ground. Frame 6 shows the Bigfoot bending over again. The pronounced dome known as the "sagittal crest" can be seen clearly in this frame. I think

this Bigfoot stayed parallel to me the entire way back to area where I capture footage of two adolescent Bigfoot, one hiding behind a tree and the other in the stream. When I was about to discover the two smaller Bigfoot the larger Bigfoot following me hit me with infrasound which caused me to lose my memory of the event and stop filming abruptly. I did not realize I had this footage until I was reviewing the video for this journal a year later.
I continue walking down the stream unaware I was being followed by the large Bigfoot. I begin to near the public area when I capture a small Bigfoot looking around a tree at me.

The above composite photograph shows the infant Bigfoot peeking around a tree as I approach.

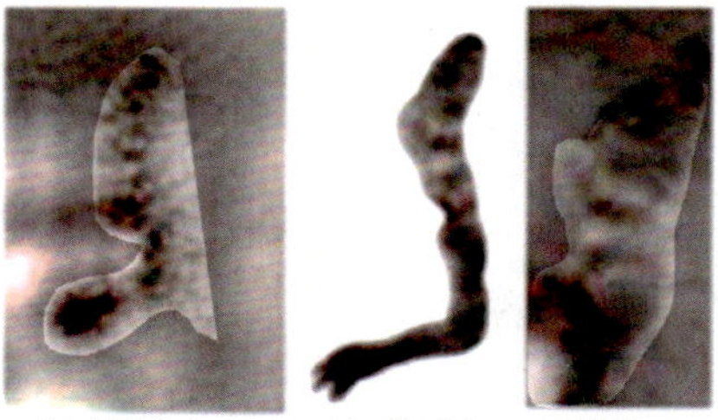

The above and below photographs show the Bigfoot peeking around the tree

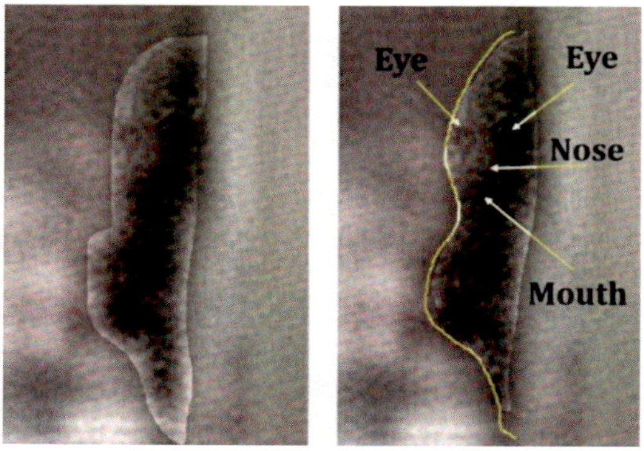

Below is a sketch using components of the actual still capture

The disturbing fact about this encounter is I do not remember it. I inexplicably stopped filming and did not start filming again for over five minutes. What happen during those five minutes? What did I see and do not remember? I think I was again hit by infrasound that caused me to lose my memory. I know this is a fantastic claim, but I did not find this footage until I was reviewing my videos a year later. The next clip contained footage of a deer I had walked up on.

I joined my family and we made our way around the rest of the loop. Wildlife was abundant including black bears which is an indicator that an area can support a population of Bigfoot. We enjoyed the rest of our day and then headed home for the evening.

Chapter 29 - Moving Trail Cameras and a "Spooky Anomaly"

Date: May 12th, 2010
Location: Research Area A

I decided to change things up so I made another late evening trip to Research Area A. I arrived at approximately 6:40 PM and went to the stick shelter to check the trail cameras and to see if there has been any activity. I find nothing had been disturbed since my last visit so I down load the trail camera and then move the smiley face and place the camera up into the tree to look down on the rocks.

I checked camera number two and then made my way over to the area where I had the mock stick structure setup. I had limited daylight so I was moving fast around the area. I did not hear anything following me but I did have the "feeling" I was being watched. Once I arrived at the site I checked the tobacco gift and the smiley face and both were untouched again.

While I was checking the area I began to get the feeling I was being watched. I felt I was being watched from the same area where I had captured video of a Bigfoot a couple weeks before. I turned and slowly filmed the area with my HD video camera. I see something that looks like a face and I zoomed in on it. I look at it for several minutes and it looks very spooky. I finally decide it is an anomaly and continued investigating the area. When I returned home and was able to look at the image closer it was even more disturbing.

I finished downloading the camera and make my way quickly out of the area. Seeing the above anomaly shook me up and I was feeling very uncomfortable. On my way out of the area I am startled by something running from under my feet. My heart jumped in my chest as the rabbit I had almost stepped on hopped down the trail. Looking back it was a funny moment but at the time I nearly soiled myself.

I made my way out to my vehicle and left the area without incident.

Chapter 30 - A Bigfoot gives me a "Gift"

Date: May 22nd, 2010
Location: Research Location B - National Park

I decided to put a second trail camera in the National Park research area. I had found a natural funnel and placed one trail camera in this area two weeks prior. I also wanted to spend some time hiking the trail where my sons and I found the footprints back in March of 2010. I arrived at the trail head parking area at 2:30 PM on a warm sunny afternoon. I made my way up the trail with a trail camera in my back pack. The first part of the hike is level for about a quarter mile, and then the trail climbs a relatively steep hill until you level out on top. Everything seemed "normal" as I made my way up the hill but once on top everything changed. The bird calls and animal activity ceased the woods went very quiet. The hair on the back of my neck stood up and I knew the Bigfoot were near and watching. I made my way to the tree where the trail camera was mounted. It was set to monitor the trail entering the funnel.

I switch out the SD cards on the trail camera then head deeper into the funnel to place the second trail camera. I work my way back into the woods to a location where the trail narrows down and there are several trees that will make good places to mount the camera. I begin to strap the camera to a tree when I hear twigs snapping in the brush approximately one hundred yards away. I stop and video the area for over five minutes but I do not see or hear anything else so I go back to work on the trail camera.

I setup the trail camera and make my way back out to the main hiking trail. I do not hear or feel anything until I get back to the trail. I begin to get the feeling I am being followed again. I move slowly on the trail because the constant twig snapping and gut feelings keep me stopping often to video the woods and look intensely for a hand, eye, nose, forehead, leg, or anything. It can be very frustrating when you know you're being followed but not able to get a good look at the Bigfoot following you.

I make it to the point where I found the footprints and experienced the infrasound back in March of this same year. I continue up the trail about seventy five yards and find a wide place in the trail where two ridge tops form a gap. I decide to stop and eat a snack and drink some water.

I sat down, took off my pack back, and enjoyed a drink of water and some granola bars I had stashed in my pack back. The birds were singing again and it was a warm and pleasant afternoon. I almost fell asleep it was such a relaxing atmosphere, but the buzzing of mosquitoes around my face quickly got my attention. The insect repellant I had applied at the beginning of my expedition was losing its effectiveness so I took out the insect repellant bottle and proceeded to reapply a generous amount. The wind shifted and I got a large whiff of the repellent. This caused me to sneeze and cough a few times.

This moment will be important because I believe it plays a role in what happens next. I then gathered my back pack and prepared to continue down the trail when I heard movement below me and the woods became quiet again. The hair on my neck stood up and the feeling of being watched again rushed over me. I had to relieve myself so I stepped off the trail about twenty feet just in case another hiker came down the trail. I continue to hear movement down the hill below me and when I finish relieving myself I begin slowly filming the area where I think I am being watched. I did see a suspicious looking anomaly while I was zoomed in at the maximum power. Closer examination of the video shows the anomaly to be a pale looking Bigfoot hiding in the thick vegetation.

The below photograph shows the Bigfoot after it had turned its head and was looking straight at me. In the insert is a blow up of the Bigfoot's face.

Below is a side by side with the actual photograph and a composite I created based on the original photograph.

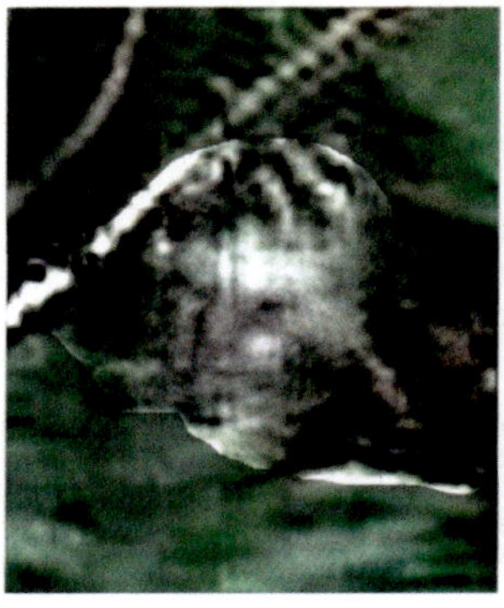

I spend a few minutes listening and watching. I am twenty five feet off the trail and hidden so I stay concealed and video in a complete circle looking for any sign of movement.

After approximately ten minutes I decide to move back to the trail and continue on my way. I begin up the narrow trail at approximately 5:00 PM. Note the time it will be important later. I traverse the trail at a leisurely pace and I film both forward and backward as I walk.

I had decided to walk out the trail for thirty minutes then turn around and walk back. This trail goes on for several miles so I kept an eye on my watch and when thirty minutes had elapsed I turned around and headed back. This is a narrow one way trail and I encountered no one on my way out the trail or on my way back. I am almost back to the location where I rested and ate my snack when I notice something amazing on a stump beside the narrow trail.

Six rocks are stacked on the edge of this stump beside the trail that I just went out an hour ago. The time is 6:04 PM. This was not present at 5:00 PM when I went out the trail. On top of the fifth rock and under the sixth is a plant leaf. I did not know it at the time but the plant that was left is Wild Mountain Mint.

I was able to confirm this by the smell of the plant which was a distinctive mint smell. I also did an internet search to confirm the plant's identity.

Notice how close the rocks are to the edge of the stump. The stack was very delicate. I decided to take the mint and leave a flower and a granola bar in return. When I barely touched the top rock the stack collapsed!! What skill did it take to stack these rocks in this manner? I also discovered that mint is used as a natural insect repellant. This raises the question; did the Bigfoot see me swatting at the mosquitoes, applying nasty smelling bug spray, and decide to give me a natural alternative? Farfetched you say? What happens next may dispel that doubt.

I replaced the mint with a yellow wild flower and a granola bar. As I am getting the bar out of my back pack I swing the camera to my left twice for only a second or two. This quick swing to my left was enough to catch a Bigfoot rising up beside a tree to watch me as I put out the granola bar.

Below is composite photograph showing the Bigfoot rising up to watch me

Before **After Head Up**

The above photographs show the Bigfoot as it holds onto the tree and watches

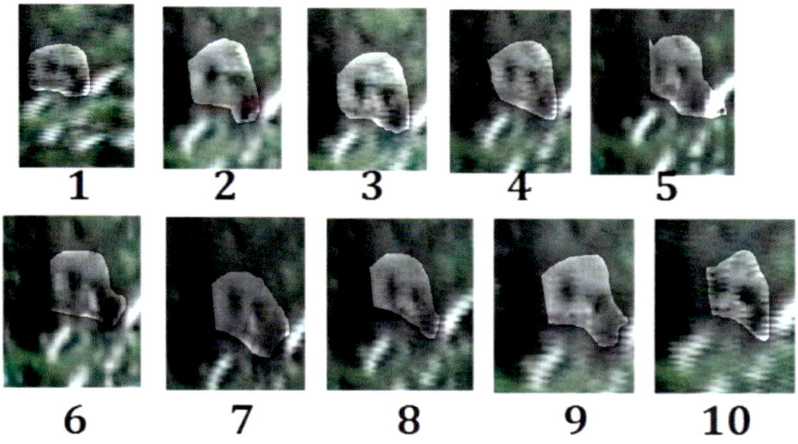

The above composite shows the one second I captured frame by frame

I did not realize this Bigfoot was watching me until after I returned home and reviewed the video. It would continue to follow me as I moved down the trail making my way down the mountain. I filmed my back trail most of the time I was in the woods this day and it paid off. Bigfoot will sneak behind you and take peeks as shown above, but they do not know which end of a camera contains the lens. So if you are not looking in their direction they assume you are not filming video in their direction as well.

I start down the trail filming behind me as I go. The Bigfoot in the previous photographs maneuvered under the hill through thick vegetation and sat down beside the trail in front of me. Then the Bigfoot went into what I describe as "stealth mode". I have a theory that the Bigfoot can use infrasound to manipulate the water vapor in the air. They "tune" the water vapor to bend light away from their bodies. This will produce a "cloaking" affect turning the Bigfoot the same color of the surrounding vegetation. I have a full chapter later in this book about this theory.

I pass within three feet of this Bigfoot and he never moves and I never see him. He is totally green and I was unaware of his presence. When I reviewed the video in which I was filming behind me I see the outline of the face, the nose, the forehead and mouth. After I pass the Bigfoot he comes out of stealth mode turns back to black and crawls across the trail and watches me as I walk away still filming my back trail.

In the above photograph I have just passed the Bigfoot in "stealth" mode. Notice it is dark green and only an about two feet off the trail.

I have returned to this trail dozens of times the "dark green" bush does NOT exist. I can find no remnants or sign that a bush was ever in this location.

05/22/2010 06/23/2011

The previous photograph shows the area one year apart. Notice the area is clear and there is not a dark green bush beside the tree.

The above photograph shows a comparison from another date after the video was captured.

I converted the photograph to black and white to reveal more details. In the following composite you can actually see the right eye and corner of the right side of the mouth.

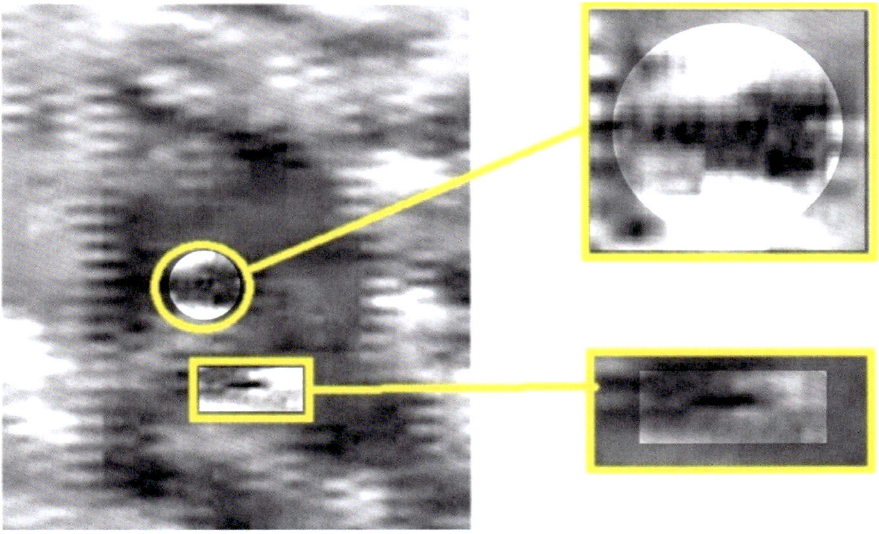

The above photographs show the right eye and the corner of the mouth while the Bigfoot is in "Stealth Mode" Below is a close up of the "eye".

I continue on down the trail about another one hundred yards when I again get hit with a very strong feeling that I am being watched. I do not know where the Bigfoot is located. I stop in the trail and begin to video in a complete circle zooming into areas that look like possible hiding places. I spend about ten minutes going very slowly and trying to be thorough as possible. When I returned home and reviewed the video my method paid off. While zooming in on an area I caught a few seconds of a Bigfoot watching me. I first thought it was possibly just an anomaly caused by shadows, leaves, and limbs, but when I went frame by frame I could actually see the eyes moving. Also during the clip the sun was reflecting off both eyes. I could also see the nose and mouth.

The above photograph shows how well the Bigfoot is concealed

The above still captures show the Bigfoot's eyes as they move and reflect light.

The next two pages of still captures show the entire sequence of the Bigfoot video. Notice how the eyes move and reflect light during the forty frame sequence.

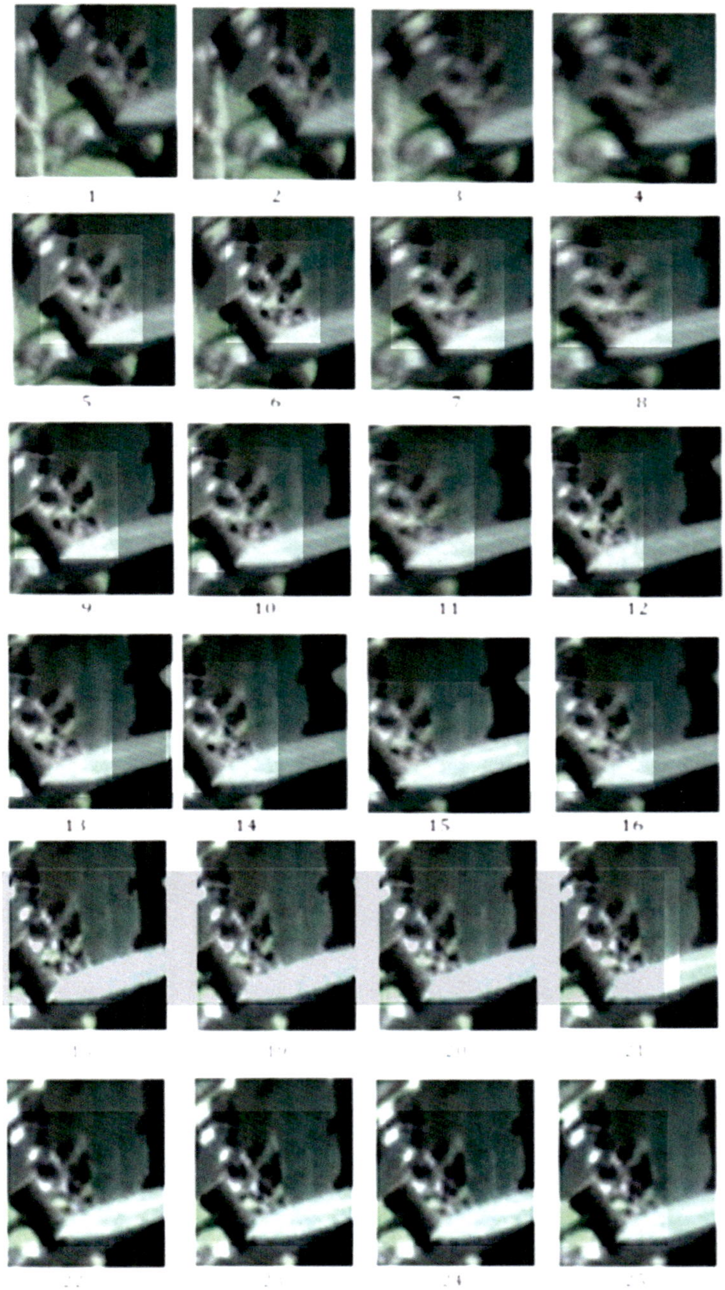

26

27

28

29

30

31

32

33
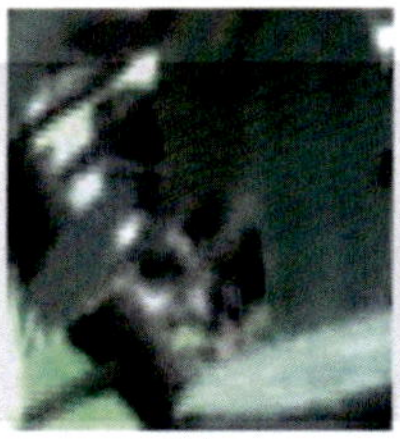
34

35

36

37

38

39

40

I used a still capture from the above make the following composite enhancement. All the elements of the composite are from the still captures. This I think creates a reasonable facsimile of what the Bigfoot looks like.

Composite with Eyes

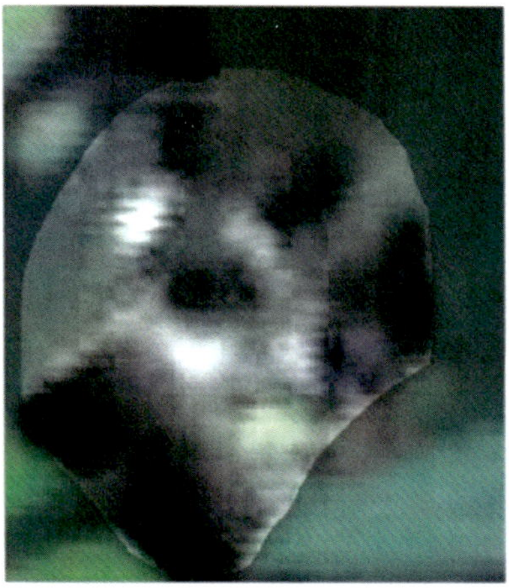

Still Capture Frame 32 of Video

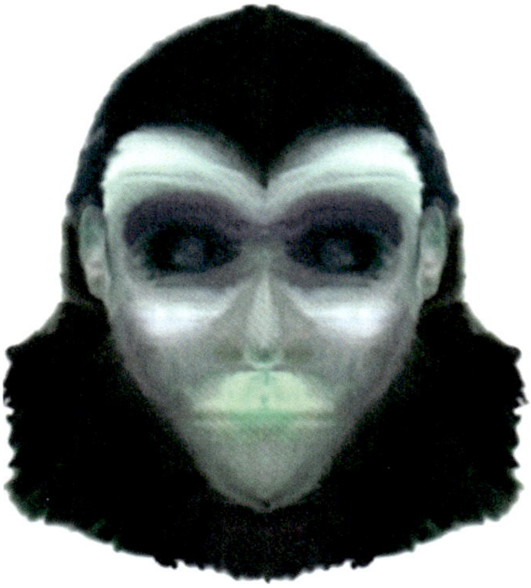

Composite Enhancement

Actual Still Capture from Video

After slowly filming in a full circle I continue walking down the trail. I hear a "beeping" noise and look down to see that the battery on my HD video camera is about to go dead. I am forced to take out my low end HD camera and use it to document the rest of my expedition. I had the feeling I was being followed as I made my way off the mountain. I came to a place in the trail where there is a sharp right hand turn. As I made the turn I heard something back to my left in the woods. I turned and looked and saw something dark and hairy climbing down out of the top of a tree. I assumed at the time it was a bear, but reviewing the video I am now unable to say for certain what was climbing down the tree. The location where I observed the "hairy" subject is a good observation point to keep tabs of what is going and coming along the trail.

Chapter 31 - Wild Mint Gift Follow-Up

Date: May 23rd, 2010
Location: Research Area B - National Park

I was so excited that a Bigfoot had left me a gift that I decided to return to the same location the next afternoon. I was hoping the Bigfoot would still be in the area and I could get more video footage.

I arrived at the parking lot in the early afternoon. I was going to check and see if the Bigfoot took the granola bar I had left the day before. My plan was to leave more food and put a hidden camera in the area. Then I would walk out the trail for approximately an hour hoping while I was gone the Bigfoot would sneak in, snatch the granola bar, and the whole event would be captured on video.

I was walking briskly up the trail because I was excited and wanted to get up on the mountain and see what had happen to the rock stack and gift overnight. I noticed a very loud bird call as I walked quickly up the trail. The call sounded out of place and it sounded artificial. The interesting thing about this "bird" is that it followed me up the trail and three miles back into the mountain. I think it was actually a Bigfoot warning others of my presence and location. I work my way back to the rock stack location and find the granola bars are missing and the rock stack has been knocked down.

The granola bars are missing, the rocks are knocked over, and the flower is missing. Did the Bigfoot do this or did an early morning hiker see this and mess with it?

I gather the rocks and stack them back up. I place a small branch with leaves under the top rock. Then I place an open granola bar on the stump. While I was stacking up the rocks I noticed a very fresh leaf lying beside the stump. I appeared that something had been munching on the leaf and took a bite out of it. The odd thing about this leaf is it does not match any of the vegetation around the stump. I wonder where it came from and what kind of plant it is.

I then set up a hidden camera a few feet away watching the rock stack and granola bar. My plan is to walk up the trail for thirty minutes. Then turn around and come back down the trail. This would give the Bigfoot an hour to come and snatch the bait. I was hoping to get them on video as they took the granola bar. I sat on the stump and while drinking water and eating a snack and I nonchalantly placed the camera down beside the stump, and slowly covered it with leaves.

Once I had the camera setup, I put my back pack on, turned on my hidden camera, and started off down the trail. Unfortunately just a few yards down the trail I met a couple coming in the other direction. Knowing they would see what I had left I explained to them that I was a Bigfoot Researcher and had an experiment setup just down the trail. I asked them to please leave it alone and told them I would remove the granola bars when I came back through. I did not tell them I had a hidden camera watching the area. The man looked at me with a suspicious grin and told me he would leave it alone. Unfortunately he was a dishonest person and my hidden camera caught him picking up the granola bar and throwing it into the woods!

I debated whether or not to even mention this but I think it is important to relate to the reader all the experiences I have had researching Bigfoot including unscrupulous people acting badly.

I make my way down the trail unaware that my bait has been thrown away and my trap ruined. I have walked for about ten minutes when I start getting the feeling I am being watched. I also began hearing the same out of place bird calls I heard when I first started on the hiking trail over three miles away. I keep walking filming to my rear and to the side most of the time. I have learned that the Bigfoot is unable to tell what direction you have the camera pointed. They watch your face and eyes to determine where you are filming. So when your back is turned or you are not looking in their direction they will peek out and take a look at you. When I film to the rear or to the sides sometimes I can get a few frames of the Bigfoot watching and following me.

When I returned to the location where I had placed the hidden camera I was very excited to see the granola bar gone. I thought I might actually have captured a Bigfoot taking the granola bar, but I also was worried that the two hikers had messed with it. When I returned home and checked the hidden camera video my worst fears were realized. The hiker reached down, picked up the granola bars and then threw them into the woods. I was flabbergasted that someone would be so unscrupulous.

Chapter 32 - Bigfoot Hiding in the Trees

Date: June 5th, 2010
Location: Research Area B - National Park

I arrive at the parking lot at approximately 9:30 AM. It is a partly cloudy day, warm, with a chance of a passing shower in the mountains. I have my rain suit packed in my back pack just in case as I gear up. I also have plastic freezer bags to cover my video cameras. I start down the trail at a normal pace and I am only about a hundred feet from the parking area when I get the distinct feeling I am being watched. I should have stopped and filmed the area but instead I swung the camera to my left and filmed as I was walking past the area. This was a mistake that I still regret. The hill side was covered in thick laurel bushes and undergrowth. When I reviewed the video I found an interesting "anomaly" that looked suspiciously like a Bigfoot but with a green face. The algorithm of the camera will affect the photograph as I have demonstrated in earlier chapters. I have decided to include the still captures in this chapter because I think this is the face of a Bigfoot. I only captured two usable frames of the Bigfoot.

The above composite photograph shows the Bigfoot standing "in" a small tree in frame one

The below photograph is a demonstration of how well the Bigfoot can camouflage themselves

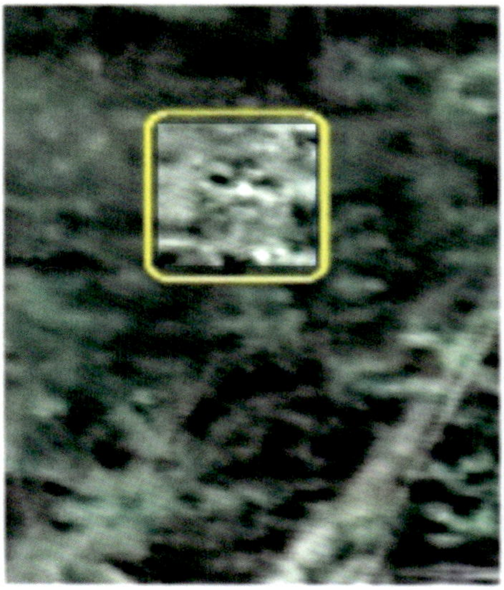

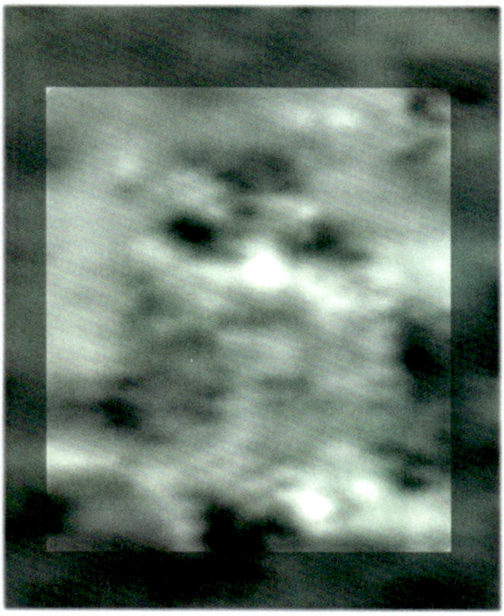

The below left is an enlargement of the actual "Face" and on the right is a composite of the Bigfoot. I have only used elements from the photograph to create the composite.

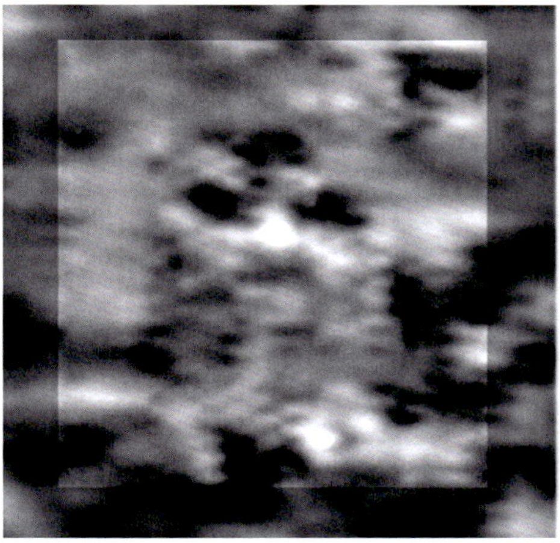

In the second frame the Bigfoot has turned its head slightly toward me removing the shadows from the eyes. In the following composite photograph the eyes of the Bigfoot are visible when the photograph is converted to black and white.

After I finish panning the hill side I continue on and climb the trail to the top. I have a trail camera placed in the "Funnel" and I am going to check it. I wait on the main trail until I am sure no one can see me then head into the woods. When I make it down to where the camera is located I discover that the SD card I used was incompatible with the camera. It was reading full but had no video clips on it. That is when I realized the camera could only accept 2 GB SD cards and I had placed a 4 GB SD card into it.

I have this camera monitoring what I call a "high head clearance" trail. This trail goes through the natural funnel. I worked my way down the stream that runs through the middle of the funnel. I was looking around the creek bank when I noticed a "Y" trail marker in a nearby tree. I have documented this type of "trail marker" at Research Area A. This is the first occurrence of this phenomenon at the National Park Research Area.

I find what appears to be a "toe slide" on the side of the stream bank. Something with five toes stepped across the small stream and slipped in the loose earth leaving toe marks. It would take a six foot stride to step across the stream. I am positive the toe slide print was not made by a bear because the shape of the foot is not right and there are no claw marks. I found a bear print in the mud close by this area and the bear print had clear claw marks. The toe slide did not have any claw marks.

I also find a couple small tree breaks that appeared since my last visit to this area approximately three weeks ago. I finished my expedition without further incidence and rejoined my family for a relaxing afternoon.

Chapter 33 - Father's Day Picnic, Bigfoot Peek-a-Boo, and a Close Up with A Bear

Date: June 19th, 2010
Location: Research Area B

We decided to celebrate Father's Day in the mountains and go to a picnic area in National Park. The area is always crowded and the picnic sites are first come first serve, so I rose early and loaded the truck with everything we would need for a day in the mountains. The menu would consist of hot dogs, hamburgers, and chips. I was looking forward to a restful day and enjoying my family. I packed my camera just in case I got an opportunity to take a short hike during the day.

We arrived at the picnic area early around 8:30 AM and found a large area with plenty of shade trees. We setup our chairs and unloaded the coolers then settled in for relaxing day.

The temperature would begin to rise quickly and soon we would all be heading down to the cold mountain stream than ran through the picnic area. We all took refreshing dip, even the dogs.

After we cool off I prepare to cook lunch. I light the charcoal and get the grill ready for the hamburgers and hotdogs. We have a great lunch and with full bellies everyone was laying around taking naps. This was my opportunity to take a short hike on the trail that started at the picnic area. I found my video camera and started up the trail. The picnic area serves as the trail head for three separate trails. I decided to follow the trail that stayed along the stream since the Bigfoot use the water sheds to navigate. I was approximately thirty minutes into my hike and was preparing to turn around when I began having the feeling I was being watched. I turned to my right and began filming an area of thick vegetation along the stream. I then had a strong sense that something was hiding high up in a nearby tree so I turned and filmed that area. I should have kept the camera focused on the vegetation along the stream, because when I reviewed the video from this day I found a Bigfoot peeking out from behind a tree watching me.

I only captured thirteen frames of this Bigfoot but a few of the frames are clear enough to see the eyes, nose, and mouth.

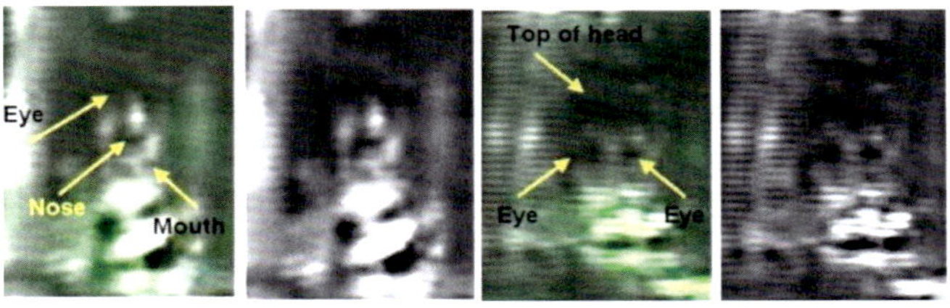

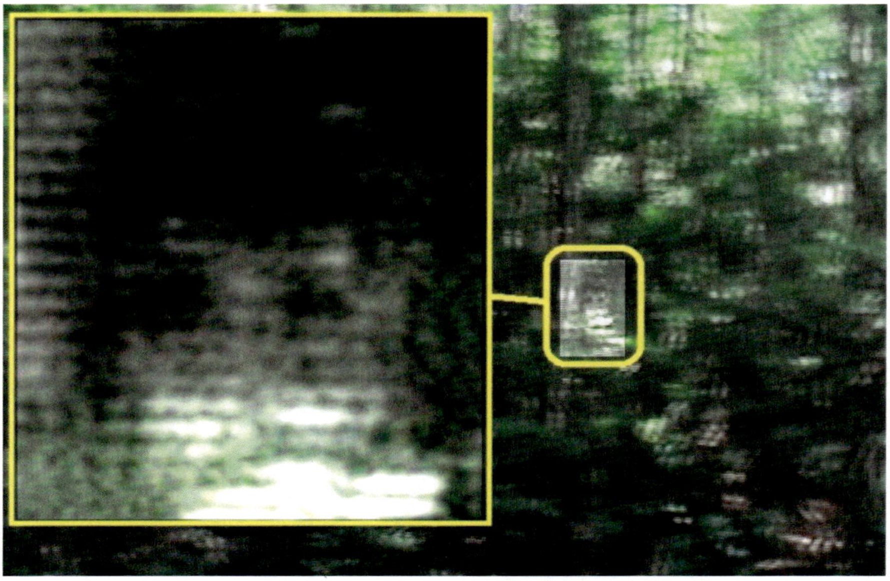

The above photograph shows the area where the Bigfoot was hiding

I returned to our picnic site and enjoyed more time with my family. We decided to wrap the trip up with a drive around the scenic loop. I would have about thirty minutes before everyone would be ready so I took my camera and headed up a second trail that originated from the picnic area. I had walked down the trail approximately fifteen minutes when I heard something large moving through the thick brush on the upper side of the trail. It was coming down the hill directly toward me. I immediately stopped and prepared myself. "Is this a Bigfoot or a Bear" I thought to myself. I readied the camera and also took out the "Bear Mace" I always carried. I did not know what was coming but I knew the encounter was going to happen fast. Then I could see the brush moving approximately twenty yards away from my location, it was a Black Bear! It was going to cross the trail directly in front of me so I slowly started to back down the trail. I did not want to surprise this bear. I backed off twenty feet, and then began trying to find the Bear with the video camera. The bear stopped short of the trail and stuck its head out and looked directly at me. The bear and I had a momentary standoff, it was looking at me and I was looking at it. Then suddenly the bear turned and made a hasty retreat back into the brush.

This encounter with the bear taught me a great deal about how quickly a Bigfoot encounter happens. I had only seconds to find the bear, zoom in, and focus. The entire encounter lasted approximately ten seconds. I only had five seconds to get the close up video.I hastily retreated and headed back to the picnic area. I told my family of the encounter with the bear and played the video for them to see. We then took a trip around the scenic loop. This ended an eventful father's day.

Chapter 34 - Watching the Trail and Footprints

Date: June 26th, 2010
Location: Research Area B

My son and I arrive at the parking area at approximately 4:00 PM. It is a hot and muggy afternoon. I have planned to check the "funnel" area for footprints and switch out the SD cards in the trail cameras that I have placed in the area. We gear up and begin walking up the heavily used trail. The first section of the trail goes between an area with very steep slopes. The area is covered in laurel thickets. Approximately one quarter mile up the trail there is a cave entrance that is at least one hundred feet above the trail. This cave is the idea observation post for the Bigfoot. They can stay concealed and monitor all the trail activity. Many times when we pass this area we will get the intense feeling that we are being watched. I always stop and zoom my video camera in on this area especially since I have already captured footage of a Bigfoot here.

When we reach the cave I stop and begin my filming the area slowly while at full zoom on the video camera.

The above composite photograph shows the entrance to the cave that overlooks the trail

I was slowly panning to the right when I noticed "eye shine". I see two eyes that appear to be reflecting the sunlight. I kept the camera focused on the area and captured the eyes moving and blinking.

The following photographs show the eye shine from the Bigfoot that was hiding in the black shadows of the cave entrance.

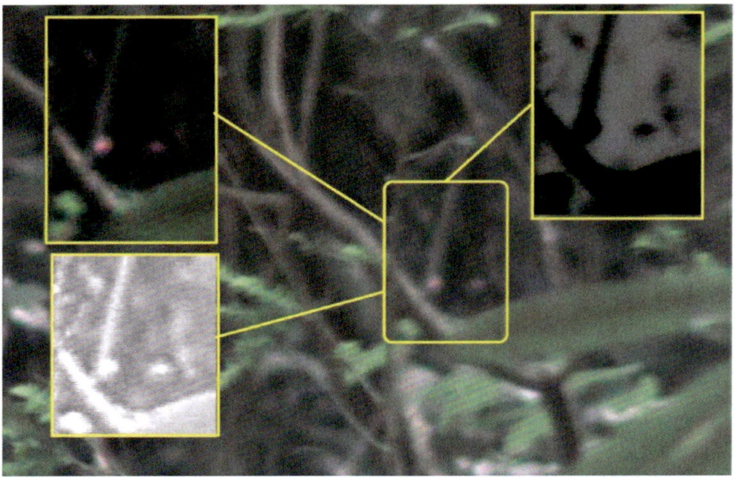

The above photographs show the Bigfoot's eyes as it watches us from the cave entrance

I try to stay on the eye shine as long as possible but the Bigfoot moves back into the cave after approximately two minutes. I pan to the left and catch a few frames of a second Bigfoot just below the area where the eye shine was located.

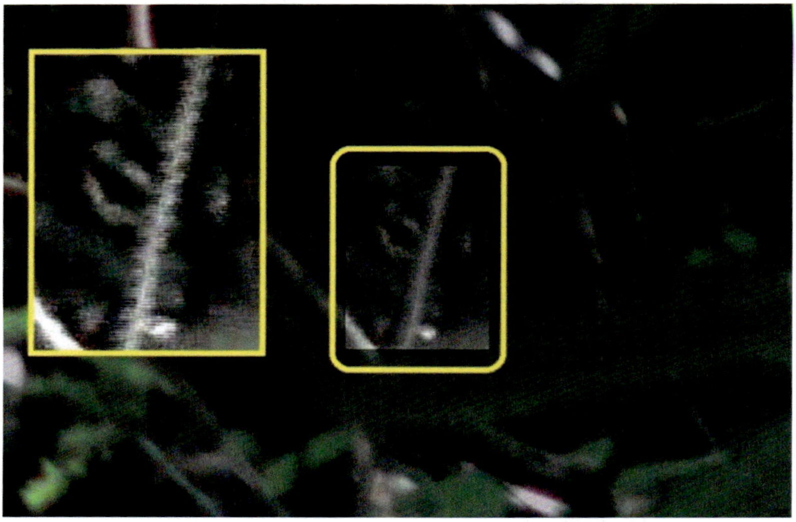

The above composite photograph shows the Bigfoot peering out of the cave

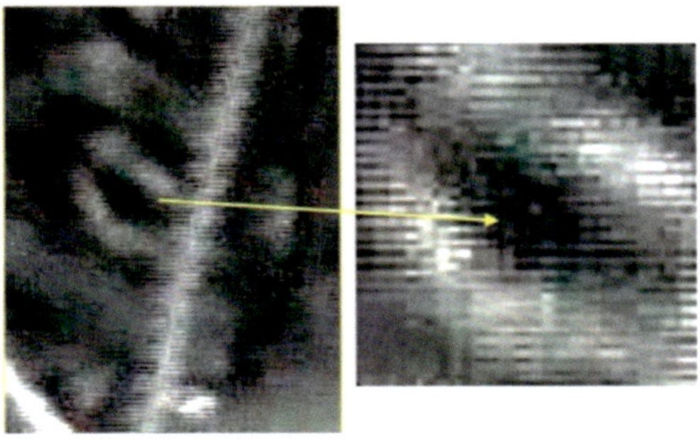

Above the eye of the Bigfoot is visible

Using the still capture I create a sketch of the possible appearance of the Bigfoot

I continue to video the cave entrance for another five minutes but I am unable to capture anymore eye shine. The Bigfoot was undoubtedly aware we were watching them and retreated into the cave. My son and I decided to resume our hike to the top of the mountain.

We reached the top without any more incidents and prepared to go off trail into the "funnel" where the game cameras where located.

We pause for a minute and make sure there are no hikers near. I do not want anyone to know I am going off trail in this location. I want to keep the area uncontaminated and I am afraid that a curious hiker may decide to follow me into this area to see why I would be going in a thick undergrowth area.

Once off the trail I begin to find the subtle signs that the Bigfoot have been moving though the funnel. I find isolated tree limb breaks at or above six feet. There is no dead fall around the limb breaks. The breaks are actually surrounded by intact limbs of longer length.

Once we were off the main hiking trail approximately fifty yards I begin to find sign that something heavy has been walking the trail leading to and from the natural funnel. I find multiple examples of large heavy limbs broken on the ground. Something of substantial weight had to step on them to break them. These were all on the trail heading into the "funnel" area. It appears that the Bigfoot travel this route on a regular basis.

We follow the small stream on our way to where I have the first trail camera located when we find a footprint. The footprint is located on the edge of the stream and appears to be approximately a week old. The Bigfoot stepped across the stream but its trailing foot left an impression. The footprint was approximately sixteen inches long by seven inches wide.

The above photograph shows the location of the sixteen inch footprint along the side of the stream

I switch out the SD cards in the first game camera and then continue to follow the stream looking for more footprints. We travel approximately fifty yards from the first track and find another set of tracks. We find a footprint in the mud on one side of the stream and directly across the stream we find a second footprint where the Bigfoot's heel imprinted the ground as it stepped across the stream. The stride was six feet long!

The above photograph shows the length of the footprint to be approximately seventeen inches

The above photograph shows the width of the front of the foot to be approximately six and one half inches wide. The heel of the foot is approximately four and a half inches wide.

The stride of the Bigfoot was six feet as it stepped across the stream. The Bigfoot only left a partial footprint where its heel landed on the other side of the stream.

The above composite photograph shows the partial footprint and the length of the stride at six feet

After documenting the footprints we made our way back to the main trail and up the mountain to the location where I was given the wild mint on top of the rock stack. I decided to make a "Feeding Station" off the trail but near the area where I was given the wild mint on top of the rock stack.

I took the rocks from the trail that were used by the Bigfoot to make the original rock stack and moved them approximately fifty yards off the trail into an area of thick vegetation that could not be spotted from the hiking trail. I found a blow down and placed the rock stack on the trunk. I also left two granola bars, one on the rock stack and one beside it. The goal is to establish a "feeding station" at this location.

After documenting the feeding station we made our way out the area without incidence.

Chapter 35 - Bigfoot Peeking from Behind the Tree

Date: July 3rd, 2010
Location: Research Area A

It was the Fourth of July weekend and I knew the lakes would be crowded with boaters so I decided to spend a very warm morning at the research area. I entered the woods at approximately 8:00 AM and the air temperature was already eighty degrees. The humidity was very high making the heat oppressive.

I enter the woods to make my way to the stick shelter. I do not have my normal Bigfoot escort this morning. The woods are "normal". No bi-pedal walking behind me or strange bird calls. It actually concerns me that I am not being followed, has something happen to the Bigfoot? Have they been harmed in any way? I now think of them as "friends" and was genuinely concerned when they did not follow me like they have in the past.

I noted the lack of Bigfoot activity and made my way down the over grown horse trail to the stick shelter area. When I arrived I noticed that the rocks had been moved around again. I inspected them and noticed that the rocks were once again in a "four – one" pattern. I had originally left five rocks on the flat piece of marble. They were arranged so each rock was touching the other. When I returned one of the rocks was not on the marble. The remaining rocks had been spread out with at least a one inch gap between each one. What was the message? There is four of us and one of you?

06/27/2010

07/03/2010

The above photograph demonstrates the "movement" of the rocks. The interesting aspect of this is how the small rock on the far right stays on the marble. If a small animal was doing this the rock would be moved off the marble slab. Also notice how uniform the gaps are between the rocks.

I also document sign of something heavy moving in the nearby area. Half inch broken limbs on the ground show something heavy walked through the area recently.

I finish up at the stick shelter location and make my way to the mock stick structure area. I am about half way to the mock stick structure when I get the feeling I am being watched. I stop and slowly video the heavy vegetation to my right. I spend approximately five minutes taking video, watching, and listening. I know something is out there, but I just cannot find it! I finally give up and keep moving on toward my destination. When I return home and review the video I find the reason I felt like I was being watched. A Bigfoot was hiding in the thick undergrowth approximately 100 feet away on my right.

The above still capture composite shows the Bigfoot and his location. The algorithm of the camera has turned the face green, but the brow around the eyes is still visible. The left ear is also recognizable in this still shot.

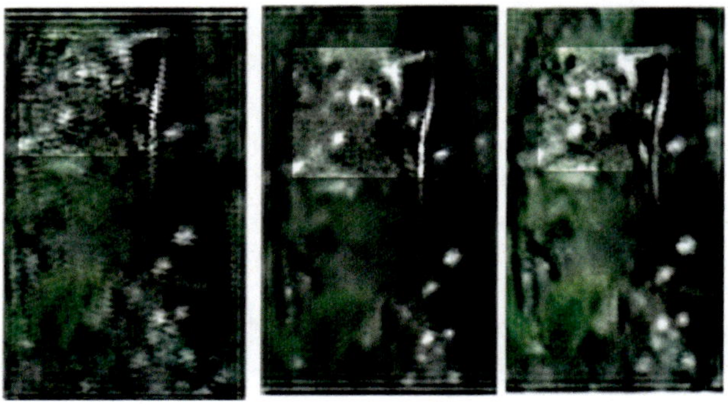

The above composite photograph shows three different frames demonstrating the Bigfoot moving its head.

In the below composite on the left photograph I point out the facial features of the Bigfoot hiding in the heavy vegetation. The right photograph is my impression of the Bigfoot.

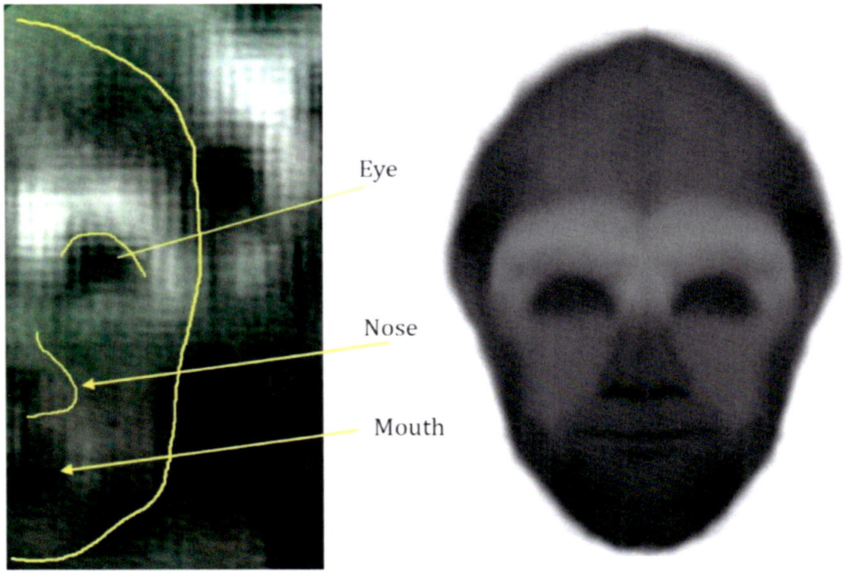

Since I did not see anything while I was in the field I moved down the horse trail to the mock stick structure location. When I arrive I notice that the tobacco gift is finally missing! Question now is what took it? Did the Bigfoot take it or was it a small animal?

I find my answer a few minutes later while looking around the area. I find the tobacco approximately fifty feet away from the stick structure on the ground. It appears it was chewed open by a small animal then left.

I decide to try something different since the Bigfoot did not seem interested in my "gift" of tobacco and placed a granola bar on the mock stick structure.

After placing the granola bar on the stick structure I began getting the feeling I was being watched. It was a familiar feeling and very strong. So I stand still and slowly pan the video camera from my left to my right. I do this for approximately fifteen minutes. While in the field I did not see anything, but upon review later I found something very interesting. I could see the forehead, brow, and nose of a Bigfoot hiding in the thick vegetation to my left. He was apparently watching me as I went about my business at the stick structure.

In the above photograph I highlight the area where the Bigfoot is hiding. Note that in the insert you can see the detail of the left eye, pronounced brow, and the nose.

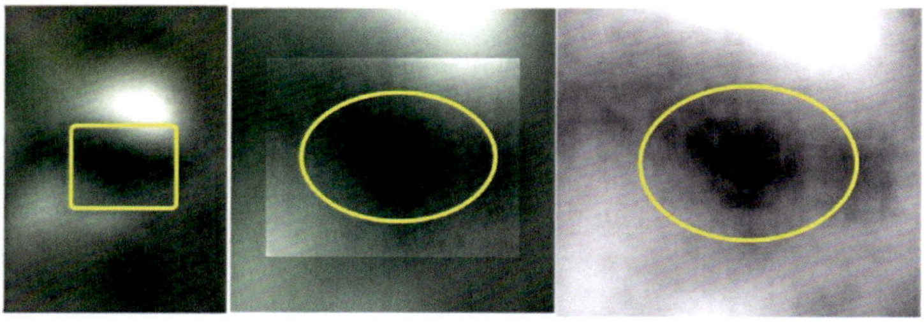

The above photographs show the Bigfoot's left eye

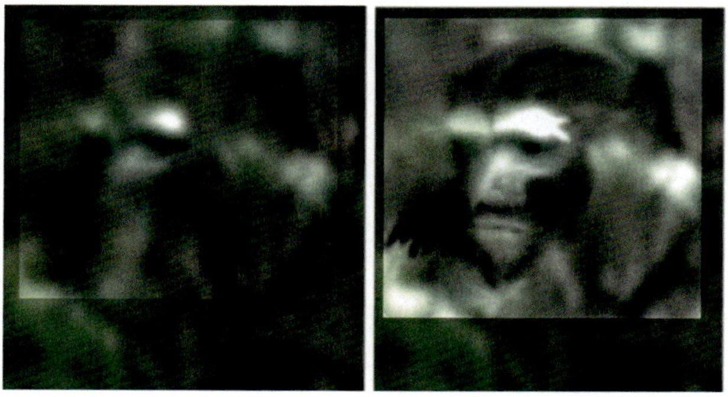

The above composite is my interpretation of the Bigfoot

I video the area but never see the Bigfoot. I leave a granola bar on the mock stick structure and work my way down to the point to check out that area. Once I reach the end of the point I find a large "bed" under a recent blown down tree.

The below photograph shows the stick structure I suspect was made by a Bigfoot. Note how a sapling was balanced in the middle of the X structure

I walked over to the blow down and notice the leaves and the vegetation pressed down and compacted under the blow down. The area was seven feet long and four feet wide.

After I document the bed I made my way off the hill an old to an old logging road. On the way in to the area I had placed duct tape sticky side out on several tree limbs hoping to get hair samples from a Bigfoot that may walk the trail after I leave the area. To my surprise I find one of the pieces of tape on the ground and it contained a single brown hair! There are no bears in this area and the tape was placed six feet off the ground on an overhanging limb. No one had been in or out of this area since I had passed through. Is this a Bigfoot hair?? I completed my trip without incident and went home for the day.

Chapter 36 - Bigfoot and Ghostly Figures

Date: July 11th, 2010
Location: Research Area A

I entered the woods this day and I knew things were different. I was being watched the moment I entered the woods. I was so overwhelmed by the feeling that I stopped and videoed an area just off of the trail for approximately five minutes. When reviewing the video I found the reason why I was so jumpy. I identify three possible Bigfoot hiding and watching me. They are just thirty feet away from where I am standing and the below still capture highlights the location of each Bigfoot.

Bigfoot number 1 cannot be seen very well from this angle but I will pan right for a few seconds then back and he will come into focus. Bigfoot number 2 is an infant lying on the ground. Bigfoot number 3 is not in focus at this time but will become visible when I pan to the right. I think Bigfoot number 1 and 3 may be in "Stealth Mode" at this moment. The longer I stand at this location the clearer they become. This hints they may be running out of breath to sustain the frequency necessary for them to stay "cloaked".

Bigfoot number 3 is the first to come into focus. I think this is a female but that is just an educated guess.

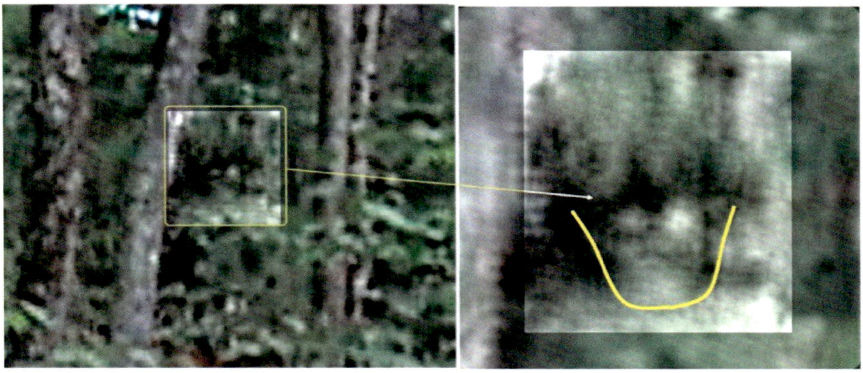

Bigfoot number 1 was hiding to the left of Bigfoot number 3. The Bigfoot was also hiding part of his face behind the heavy foliage. I think this is the "Big Male", again this is an educated guess.

Bigfoot number 2 was lying on the ground in front of Bigfoot number 3. I think this one is the infant based on size.

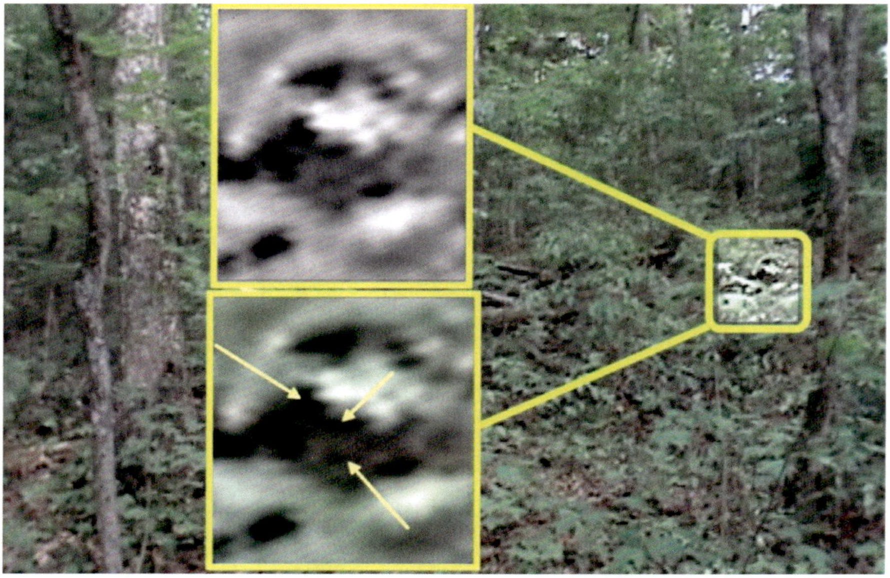

Bigfoot Number 1 – Alpha Male

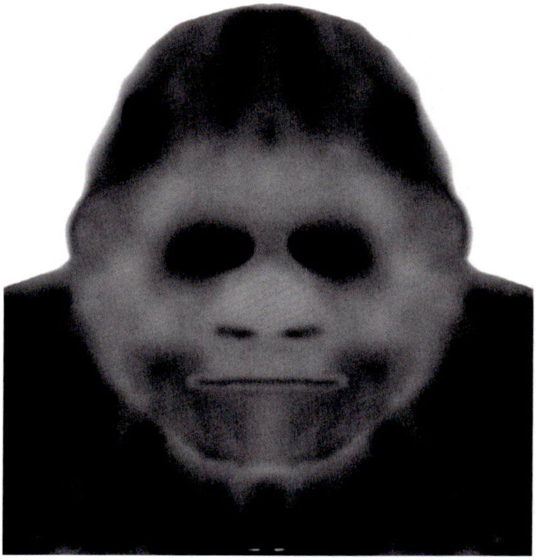

Bigfoot Number 2 – Infant

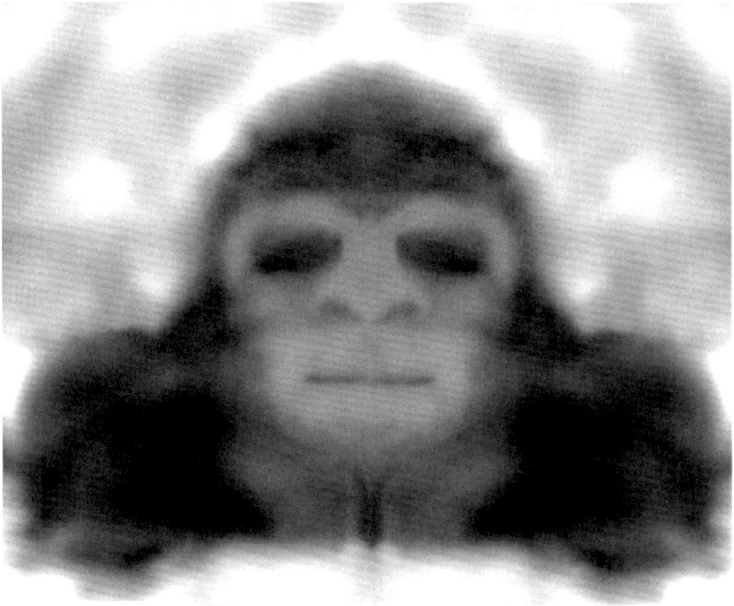

Bigfoot Number 3 – Female

I video this area for approximately five minutes then move down the horse trail. I go approximately one hundred feet when I capture the first "ghostly" figure. I capture a white colored face peeking above a blown down tree.

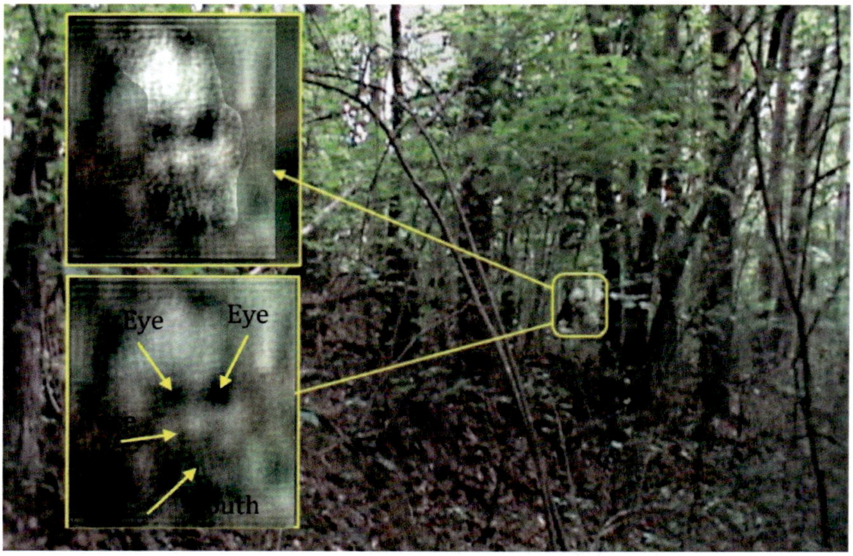

This possible Bigfoot has a high forehead but a very human looking face. Is this a person just hiding from me? Is this a feral human that is living with this Bigfoot clan? Is this a hybrid, a mix of a "human" and a Bigfoot?

The below composite shows the actual still capture and a composite enhancement. The subject looks very human!

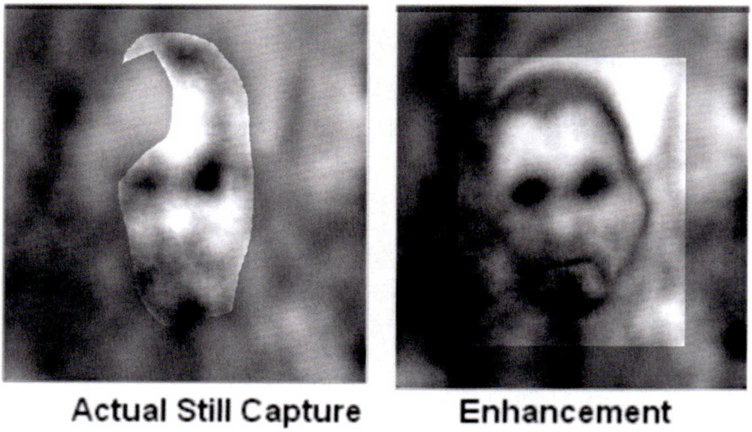

I walked past the above "figure" and was unaware that it was that close. I knew I was being watched but did not see the subject until I reviewed the video. I cross the dry creek bed and continue walking down the horse trail towards the stick shelter location. I am filming to my left and right as I am looking forward. This explains why many times I never see the subjects looking at me. I am concentrating on walking and not falling on the rocks and sticks that litter the narrow trail.
There are two figures in the next still capture, one that looks "ghostly" and another that is out of focus because I am moving.

I have no explanation for this subject. It is a video an anomaly? Is it a Bigfoot in stealth mode? Is it something else? Again I paused in this area for a few moments because I had the feeling I was being watched. I find it interesting that every time I have paused and looked around I have later found "subjects" in the video.

Just a few feet in front of the above figure is something else that resembles a Bigfoot hiding in the thick foliage. I am moving the camera during most of this sequence so the images are blurry and out of focus.

I stop right beside this Bigfoot in stealth mode and look around but I never pan the camera to my left. I look forward and then to the right.

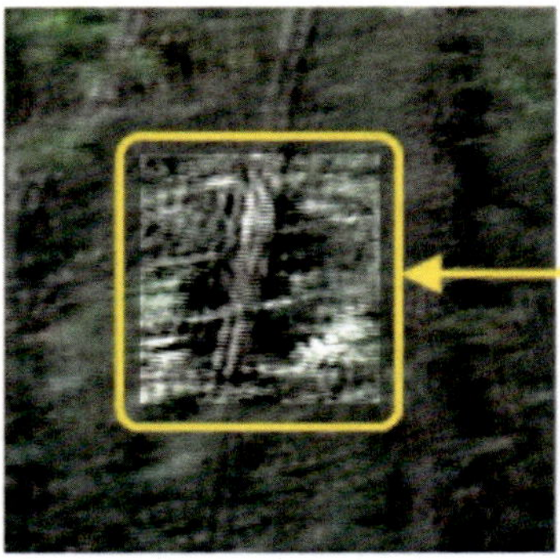

Bigfoot visiable as I pan right

Bigfoot in stealth mode as I pan back to the left 15 seconds later

In the below photograph I took the liberty of doing a composite enhancement of the Bigfoot beside the trail

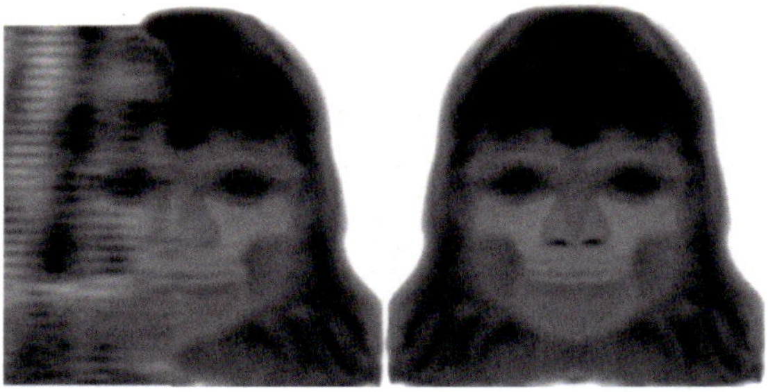

Again, not seeing any of this with the naked eye, I continue down the horse trail. I know I am being closely followed but I continue about my business of documenting the stick shelter area. Nothing had been disturbed so I moved on down the trail without incident to the area where I had the mock stick structure set up.

I place a granola bar on my mock stick structure when I began hearing movement in the brush. The movement is loud and it starts coming from several directions. I then start hearing wood knocks that are close. I finally have had enough and begin talking to the Bigfoot. I ask them to come out and stop playing games! That is when I get a strong feeling that a Bigfoot is near. I train my camera on the area and video without moving for over five minutes. I strain my eyes trying to find what has drawn my attention to the area. I video this area for over ten minutes moving the camera a few degrees right then left. When I reviewed the video I found a Bigfoot hiding in the thick vegetation.

I film this area for fifteen minutes as the Bigfoot continues to make noises and tree knocks. The commotion finally dies down and I work my way to the end of the point of land. When I arrive at the point I feel eyes watching me from behind. I slowly turn around and pan the area for approximately two minutes. I did not see this Bigfoot until I reviewed the video. The below still shot shows the small, round headed, Bigfoot.

Note that the Bigfoot is holding a plant limb in front of its face. The light colored underside of the green leaves is pointed toward me.

The eye of the Bigfoot was also visible.

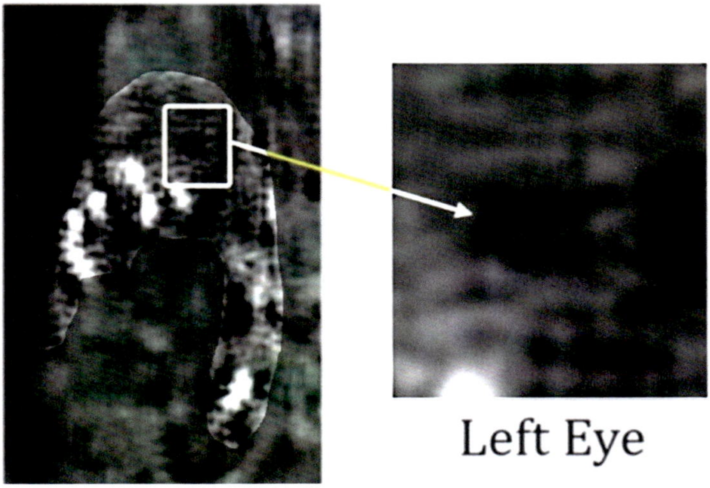

Below is a photograph demonstrating the head of the Bigfoot tilting back and forth during the video.

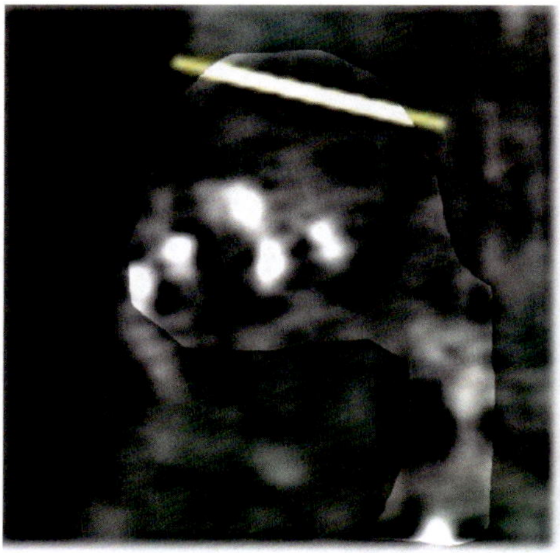

Frame 108

Frame 409

Below is a composite sketch of the Bigfoot I created using the actual photograph.

Chapter 37 – Bigfoot in the Laurel Thicket

Date: July 18th, 2010
Location: Research Area B

I arrive early to the National Park location. I have had the trail cameras out for three weeks and I am anxious to see if I have captured any video of a Bigfoot. I hurriedly go up the trail in anticipation of what I might find. The cameras are located in a natural corridor in the mountains. I have to climb a modest incline to reach the top of the mountain and the area where my trail cameras are located. I know from finding footprints and tree breaks in this area that the Bigfoot are traveling through this area on a regular basis.

I have four trail cameras placed strategically in this area. Since this area is narrow it is easier to place the trail cameras so they cannot be avoided. I know the Bigfoot can hear the Passive Infrared (PIR) sensors that trigger the camera. In order for the Bigfoot to avoid my cameras they would need to walk a route that is more difficult and time consuming. When I arrived in the area there were a number of freshly broken saplings and small uprooted trees strewn all around the area. This activity has taken place after I placed the cameras in this location. I wonder if I have angered the Bigfoot and this is their way of showing it.

I was investigating the area when I decided to take a short break and relieve my bladder. I always point my video camera to watch my blind side. So while I am relieving myself I am pointing the camera at a laurel thicket located behind me. I was unaware of the presence of the Bigfoot that was hiding behind me in the thicket. During review of the video I found it looking at me.

The below still capture shows the Bigfoot hiding with a close up in the insert, note that you can see the detail of the left eye

I was able to capture seventeen seconds of the Bigfoot watching me. During this time it is moving its mouth and eyes. The following collage demonstrates the eye movement.

Eye Squinting

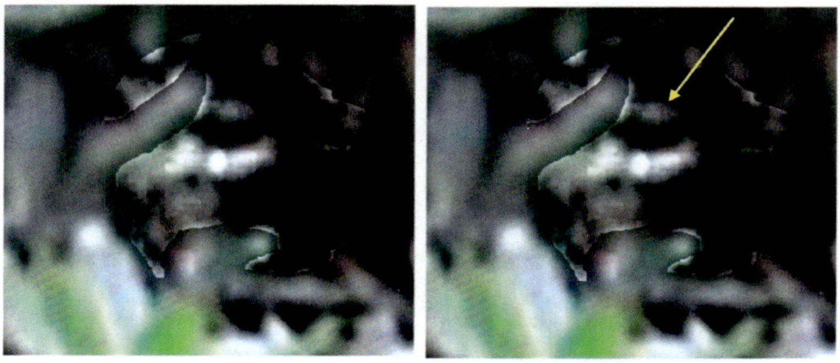

Eye Wide Open

The Bigfoot was also constantly moving its mouth. The following photograph demonstrates an instance during the video clip where the Bigfoot was moving its mouth. The photographs are from sequential frames.

Frame 1 Mouth Closed

Frame 2 Mouth Open

Later in the video the eye is visible again

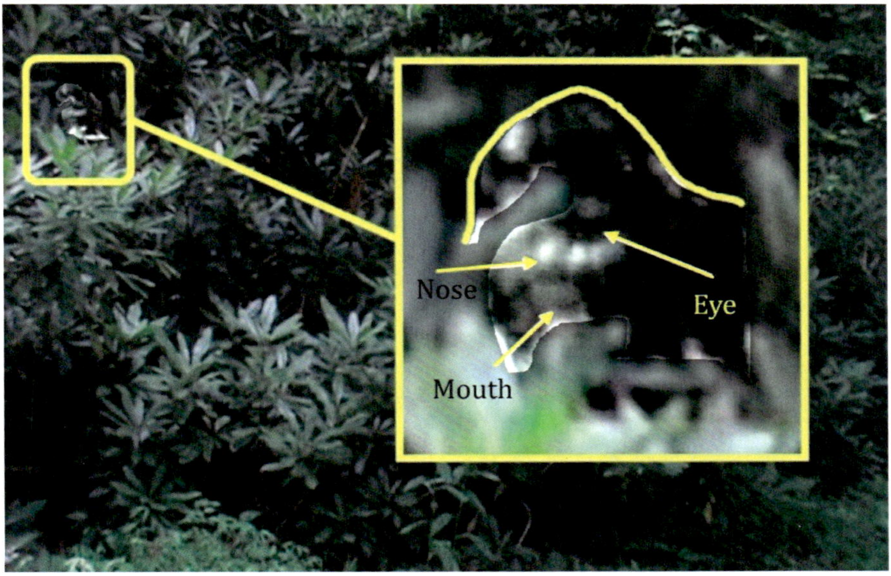

Below is a close up of the eye

I returned to the location and performed extensive follow-up. I determined the approximate size of the head and the height of the Bigfoot. A storm had blown down trees into the thicket but I was able to locate the location where the Bigfoot was standing. Below is a scale height comparison. The Bigfoot was approximately seven to seven and one half feet tall.

The head of the Bigfoot was approximately nine inches wide. The average width of the laurel leafs in this area was three inches. I used this average to determine head width.

Below is a side by side comparison photograph of the Bigfoot to me standing in the same location. Using common land marks I have scaled the pictures to match.

When I finish relieving myself I swing the camera slightly and start to take a step. When I do this the Bigfoot pulls back into the vegetation, when this happens the entire face is visible.

I work my way down the small stream that runs through the corridor looking for footprints and changing out the SD cards in the trail cameras. When I reach the far end where my last camera is located I notice a very fresh and clear print of the front part of a foot. The Bigfoot stepped across the creek and pushed off and slid in the mud.

I made sure there were no claw marks so I could rule out a bear. I could find none, nor could I find anymore prints. It appears the Bigfoot stepped across into the leaves on the other side of the stream.

The above footprint is evidence that the Bigfoot are using this area and my trail cameras are located in the correct area to capture video of a Bigfoot.
I work my way deep into the woods following the stream as it enters a laurel thicket. I come upon a pine tree that has been snapped off at the ten foot level. There are no fallen trees or limbs around it that could have fallen and broken it. There is also no other trees near it damaged or broken from a possible storm or heavy winds. It appears to be a “trail marker” left by the Bigfoot.

I investigated the area thoroughly and found no other evidence so I began to make my way back out to the main trail. On my way back out I find a pile of fresh scat. It did not resemble bear scat, but it looked like very large human feces. It had seeds and small bones in it. This is most likely Bigfoot scat, but I am unable to say for sure.

I move down the other side of the stream and find a skid mark on the ground. This could be where a footfall landed after a Bigfoot stepped across the stream and slid in the loose soil.

I work my way out to the main trail then head to the location where the Bigfoot gave me the gift of the wild mint. I have established a feeding station near this location. Three weeks ago I left a rock formation and a granola bar. I wanted to check it and see if my food was taken. The food was missing.

Below is an unedited cut out of the Bigfoot that was hiding in the laurel thicket. The photograph has not been altered.

.

Chapter 38 - Bigfoot goes for a swim

Date: July 30th, 2010
Location: Research Area A

I had prior engagements in the morning so I was not able to go to the research area until late in the afternoon. The weather was oppressively hot with high humidity. The cicadas were also extremely loud as I entered the woods. My plan was to retrieve my trail cameras so I could move them to the corridor area in Research Area B.

I continue quickly down the trail since it is late in the day and I only have a limited amount of day light left. Once I arrive at the stick shelter location I prop my camera up against a tree to film the woods on my blind side and began to take down the trail cameras. While I am working on the trail cameras I begin to feel the now familiar feeling that I am being watched. Since I have my camera pointing in the direction I am getting the feelings from I just continue about my work. I hope by ignoring the Bigfoot I can catch it moving on the camera I have pointed in that direction. Unfortunately as usually happens in Bigfoot research, the Bigfoot sneaks up on me to the right of where I have my camera aimed.

Once I finish up taking down the trail camera and placing it in my back pack I look over to the area where I feel like the Bigfoot is hiding. I see what looks like an eye peering at me out of the thick foliage. I grab the video camera quickly and point it in that direction. In the below photograph I point out the area where I see the eye moments before I attempt to zoom in with the video camera.

I then make a “rookie” mistake. I begin to zoom in with my video camera to the area where I see this eye looking at me. The closer I zoom in the clearer the eye and the face of a Bigfoot becomes! My heart is now racing and adrenaline is flowing. Finally I think to myself, I can see a Bigfoot clearly! Then I do it, I become so excited that I look over the top of the video camera at the Bigfoot instead of looking though the camera. When I look at the Bigfoot he turns his head and with both eyes now looks directly at me.

I am approximately thirty feet away from a Bigfoot, making direct eye contact. The Bigfoot is squatting and yet his head is still five feet off the ground. Unfortunately while I am making eye contact with the Bigfoot I kept zooming the video camera and it is capturing nothing but a large blur of foliage.

I was able to capture just a few frames of the Bigfoot before I took my eyes off the view screen of the camera. The below photograph show what I saw right before I looked up and made direct eye contact.

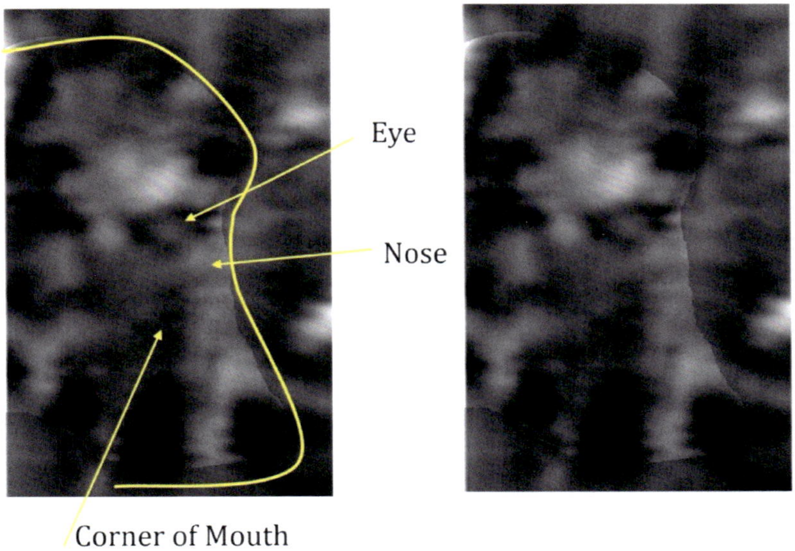

The Bigfoot and I look at each other for a few seconds. He had dark black eyes but they were not scary or ominous. The Bigfoot had a scowl on his face when we first made contact. Then the expression changed. The face relaxed and the expression changed. I think he was just curious and that he actually recognized me. It is almost like when we made eye contact the facial expression was "Oh, I know you". Then without warning he turned to his right and disappeared into the brush on all fours.

The Bigfoot retreated in the direction of the small island that was just off the main land. This island is where I had previously captured video of a Bigfoot sneering at me.

I quickly went over to where I spotted the Bigfoot and found broken limbs and sign on the ground. I decided it would not be wise to follow the Bigfoot into the thick foliage so I went back out to the main trail and headed toward the island.

Once I arrive at the shore I began looking around for any sign the Bigfoot had crossed over to the island. I looked around for approximately two or three minutes when I heard loud splashing or sloshing as if something large was walking through or coming out of the water. I could not see where the Bigfoot was but I could hear it coming out of the water and stepping onto the island.

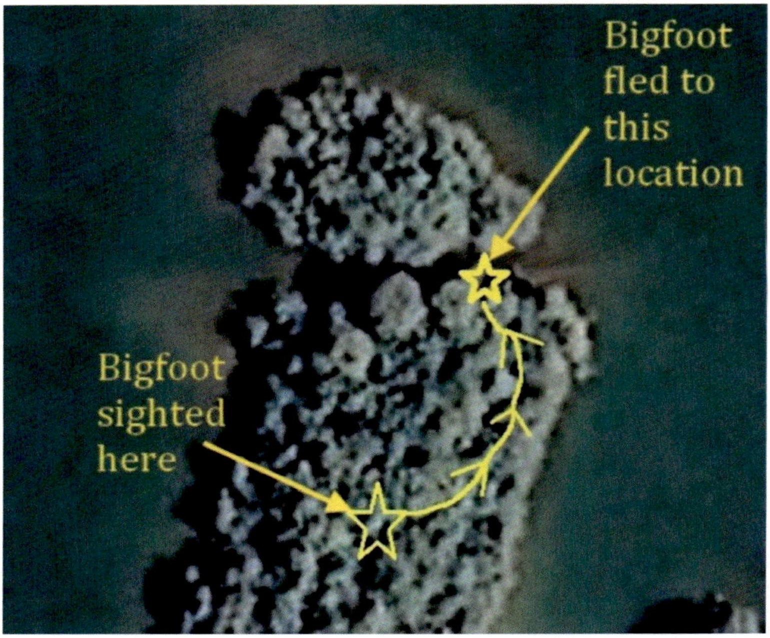

The proximity of the Bigfoot to my location was very close, no more than twenty feet. It shook me up because I was not expecting it. I must have come too close for comfort and it slipped into the water and crossed over to the island. It took me a few minutes to regain my composure. I contemplated wading across to the island but with darkness closing in and being alone I did not want to push my luck.

I made my way out to my vehicle and returned home without any further incident.

Chapter 39 - Setting the Trap

Date: August 1st, 2010
Location: Research Area B

I decided to set a "trap" for the Bigfoot in the National Park location I call the "Corridor". I continued to find footprints and other sign that the Bigfoot were passing through this area on a regular basis. I developed a plan to arrange the trail cameras to create the appearance that I left one area of the corridor uncovered, as to entice the Bigfoot to walk through this area. I retrieved all my cameras from Research Area A and enlisted the assistance of my oldest son to help me set up all the cameras.

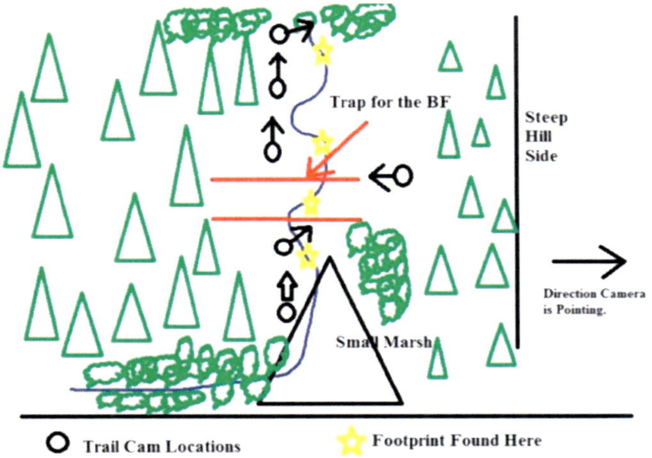

The above diagram shows how I had the trail cameras setup. The black circles show the location of the trail cameras. The arrows are pointing in the direction the cameras were aimed. The yellow stars show locations along the stream where I have found footprints in the past. I have marked in red the area where I purposely left a coverage gap. This gap is where my trap is located. The plan was to leave this gap so the Bigfoot would cross the stream at this location. I set a camera directly across from the gap to capture video as the Bigfoot crossed through. I also placed the cameras back to back so the Bigfoot could not come up from behind the cameras and disable them as they had in the past. I was also hoping that by flooding the area with cameras the multiple PIR sensor signals would confuse the Bigfoot and mask the location of the cameras. It would only take one Bigfoot that was unable to locate a camera to get the footage of a lifetime. I arrived at the location with my remaining three cameras that I had taken from Research Area A. I went to work setting up the cameras and sent my son to walk around me filming the area. I know the Bigfoot always follow and watch me when I am in this area. I was hoping that my son's presence would distract them and they would pay attention to him moving around while I mounted the trail cameras on the trees.

I used heavy steal boxes that I constructed to both protect and reduce the electronic emissions from the cameras. I also cable locked the cameras on the trees to prevent bears from pulling then off the tree. The camera boxes were completely camouflaged and painted to match the bark of the tree. They were also what I call "seasoned" cameras. This means they had been out in field for over a year and had very little if any human odor on them.

After I mounted all the trail cameras we began looking around the area for footprints and other sign. My son spotted an old footprint left in the mud on the side of the stream. The footprint was old but the outline was still recognizable.

The following composite photograph shows the footprint. It is fifteen inches long and seven inches wide at the ball of the foot.

We continue to exam the stream bed and discover that a Bigfoot has walked in the stream within the past twenty four hours leaving two impressions.

The Bigfoot are now walking in the stream bed. I would assume it is to prevent me from finding their footprints. We move on from the camera trap area to the feeding station. The food I left is missing but it is difficult to tell if it was taken by a Bigfoot or an animal.

Chapter 40 – Activity is Slow in the Heat of Summer

Date: August 14th and 21st, 2010
Location: Research Area A

My plan was to give the camera trap I had set in Research Area B two weeks. I visited Research Area A two weeks in a row. I found little activity during this period. I decided to start leaving apples for the Bigfoot at both research locations. I want to begin to build trust with the Bigfoot. The primary way to do this would be through food gifts. I placed an apple on the mock stick structure. I knew the Bigfoot visit this location so I took one bite out of the apple in case they were watching me and placed the apple in the center of the structure. Taking a bite out of the apple demonstrates to the Bigfoot that the food is safe to eat.

After I finish in this area I move to the stick shelter area. I continue to monitor how the Bigfoot move the rocks around trying to make sense of what it all means. I leave four rocks on the flat rock and come back a week later. When I return I find three rocks have been pushed together, a new rock has been added, and the small rock was moved off the flat rock.

08/14/2011

08/21/2011

Note in the above photograph on the right how the three rocks are pushed together and a larger fourth rock is added. This continues the familiar pattern of 4 – 1. The small rock is removed.

I think the meaning is still "there are three of us and one of you." The major difference is the lone rock is not that far from the group as in the beginning. I am hopeful this means that I am being accepted by the Bigfoot clan.

This is all the significant activity that happened in the two weeks between August fourteenth and August twenty first. I am looking forward to going into Research Area B to check my trail camera trap.

Chapter 41 - The Trap Pays Off

Date: August 28th, 2010
Location: Research Area B

It has been two weeks since I saturated the "corridor" with trail cameras. It is time to go in and see if the trap worked. I take my middle son who is thirteen years old at the time. We arrive at the parking area early. I have a fresh supply of batteries and SD memory cards for the trail cameras. We put our packs on our backs and head out the trail. We come to the area in the trail where in the past I have captured footage of a Bigfoot hiding on the hill side. This position is less than one hundred yards from the parking lot. I am talking to my son and pointing out a tree that had recently fallen across the trail. I am looking down but holding the camera where it is pointing up into the base of the fallen tree. My son and I are unaware that a Bigfoot has raised its head above the fallen tree and is watching us.

The above composite photograph shows the Bigfoot peeking over the top of the fallen tree

My son and I actually talked to each other about the thick vegetation that is in this area and how easy it would be for a Bigfoot to hide in it. We make our way up the trail to the "corridor" where I have the trail camera trap setup. I begin checking the trail cameras one by one. The first camera I checked had a dead battery. This was very odd. It only had two "white out" videos on the SD card. Dead batteries are a common phenomenon when dealing with Bigfoot. I switch out the SD card and replace the batteries and go to the next trail camera.

The second camera has been pushed to the base of the tree. When I check the SD card it yields no answers, just video of the ground once it was laying on the ground.

The above photograph shows the second camera on the ground

We move to the third camera and it is the most curious. The wire pine limbs that I used to help camouflage the camera have been moved in front of the lens of the camera. The SD card for this camera only contains one "white out" video.

The above photograph shows camera three with the lens blocked

The other three cameras are untouched and the SD cards contain videos. It appears that the Bigfoot were trying to disable the cameras so they could walk through the below corridor unnoticed.

The cameras they disabled would allow them to pass through this area without being captured on video. The Bigfoot again shows their remarkable intelligence. I set a trap for them and they simply disabled the cameras to allow themselves to pass undetected. I wondered why they only disabled three cameras, why not all of them? Then it came to me, because they did not need to disable them all. They only needed a hole in the coverage to walk through. Also I think that they are intelligent enough to know that their presence would be obvious if they disabled all my cameras. Just disabling the three would still leave doubt to the cause.

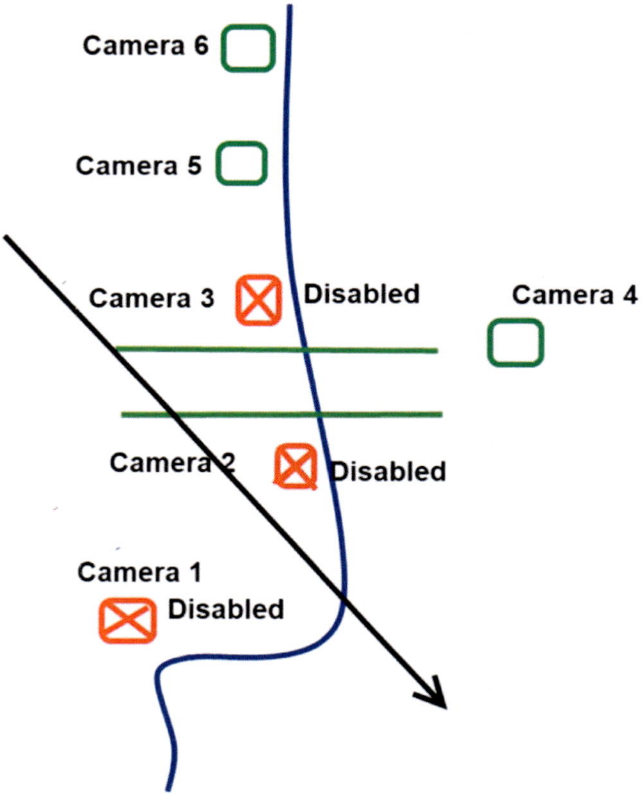

The above diagram shows the cameras that the Bigfoot disabled and the black arrow represents the path they can now follow. The two green lines show where I was hoping to lead them so I could get a video with camera number 4.

When I checked the SD card for camera number 4 it had 18 video clips and all had only footage of the woods, no Bigfoot. After checking camera number 4 we move on to camera number 5. This is where it gets very interesting. I down loaded camera number 5 and there were 3 video clips on the camera. The third and final clip was the most interesting. The camera was tripped on August 13th at 22:39 PM. In the first few frames there is something very large standing directly in front of the camera. The image had no detail because the subject was directly in front of the camera. The infrared reflected directly back into the lens so the subject is completely white. Fortunately the subject was standing where only half its body was in front of the camera. This gave a very good profile of the subject. In my opinion the subject is a very large Bigfoot that decided to check out the game camera.

The Bigfoot was rocking back and forth in front of the camera. I captured 23 frames of the large Bigfoot. The below picture is the first frame when the camera was tripped.

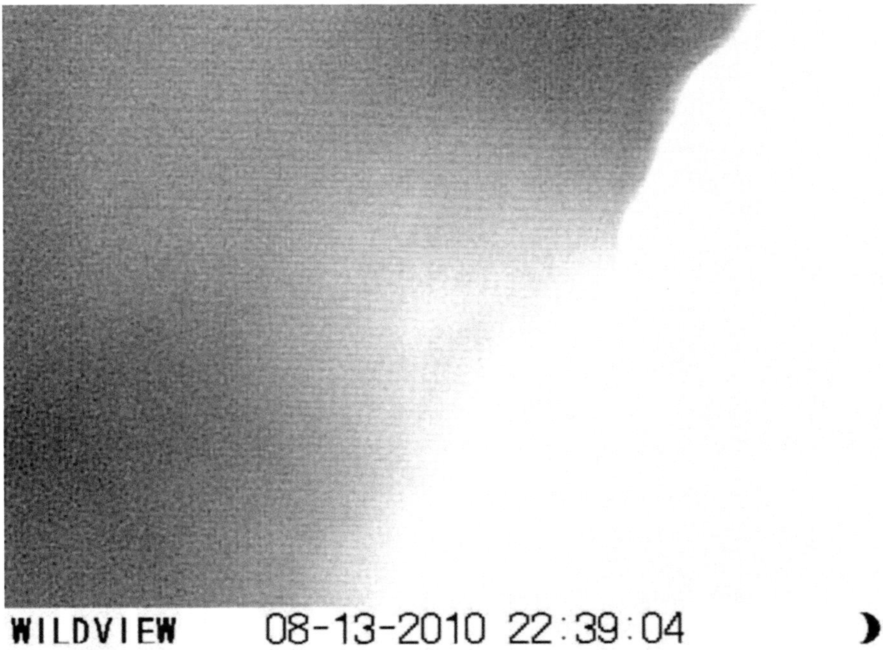

The camera was mounted on the tree at approximately nine feet off the ground. When I checked the camera it was tripped and took video of me. This gave me an excellent reference for the size of the Bigfoot since the camera was in the exact location on the tree when it captured the footage of the Bigfoot.

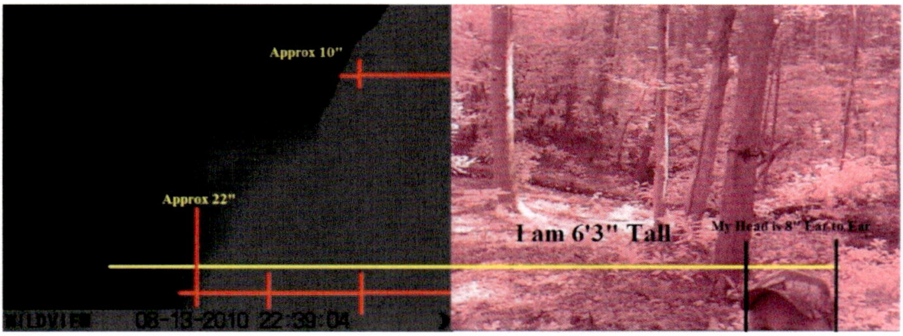

The above composite photograph shows my head with the size measurements. My head is eight inches across and I am six feet three inches tall. Using this as the base line I was able to determine the width of the shoulders, head, and total height.

Using the still capture as a reference from my eyes to the top of my head is five inches. The top of my head to the top of the picture frame is fifty inches. Add that to my height and that gives us the approximate height of the Bigfoot at over 10 feet tall!

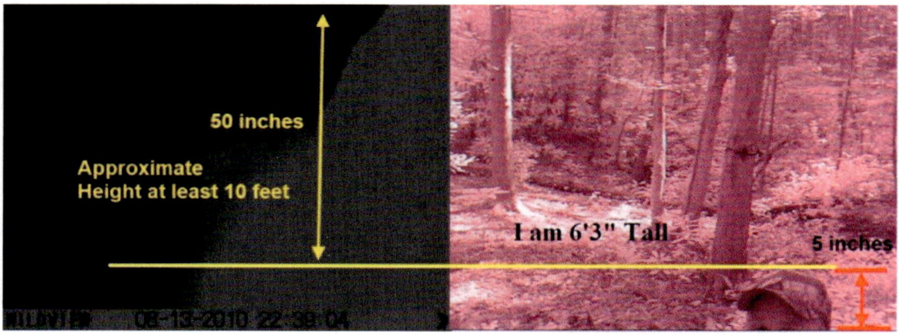

I used the width of my head to determine the total with of the Bigfoot shoulder to shoulder and the width of its head. My head is eight inches across. If we make the assumption that approximately half the Bigfoot was visible in the frame then the visible half is approximately twenty two inches. Double this and you have a shoulder to shoulder length of forty four inches. This would make the width of the head approximately twenty inches wide. This would put the dimensions of this Bigfoot at over ten feet tall with a chest measurement in excess of eighty eight inches. The face would be twenty inches across. This is one huge Bigfoot! In the below photograph I flip and copy one half of the frame to give us an idea of what the full profile might look like. The shape is consistent with many Bigfoot descriptions including the "no" or "little visible neck". The head sits on the shoulders.

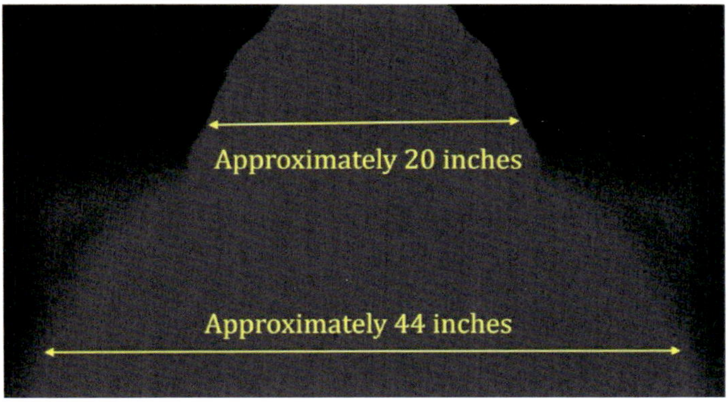

After the Bigfoot stepped to the side of the camera it appeared that this was all the action in the video. I watched the video several times and began to notice something moving back and forth between two trees in the back ground. Whatever was moving was just out of range of the infrared light. The below still capture is the area where I see what appears to be a Bigfoot peeking at the camera.

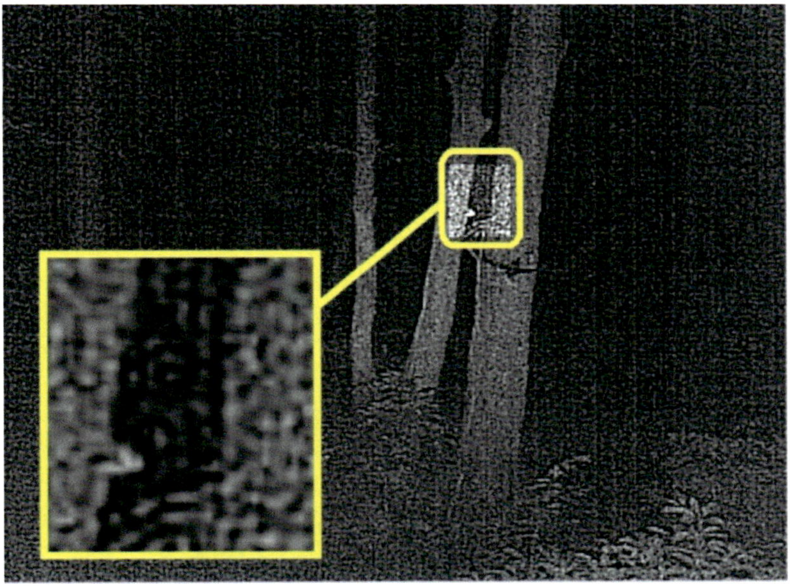

I started going frame by frame and noticed that there were two Bigfoot peeking between the trees at the trail camera. Unfortunately they were at the edge of the infrared emitters range. This would cause them to go in and out of focus as they rocked back and forth and jostled with each other for position. I realized as I went frame by frame that there were at least four different Bigfoot pushing and shoving each other for position so they could look at the trail camera.

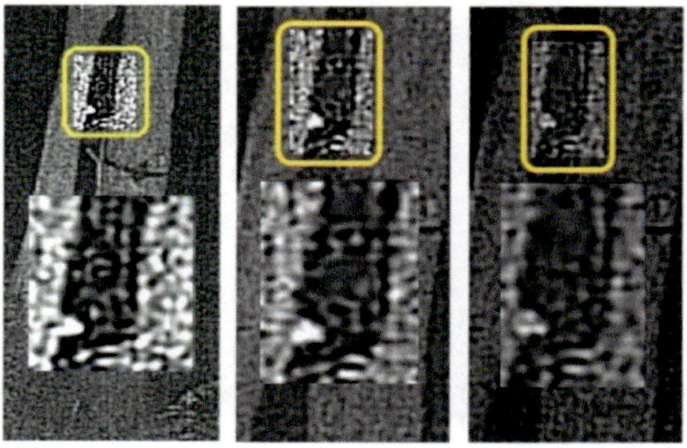

The above composite photograph shows the different frames in the video where I have identified these four individuals peeking between the trees.

The first Bigfoot is very primal looking he is approximately seven feet tall

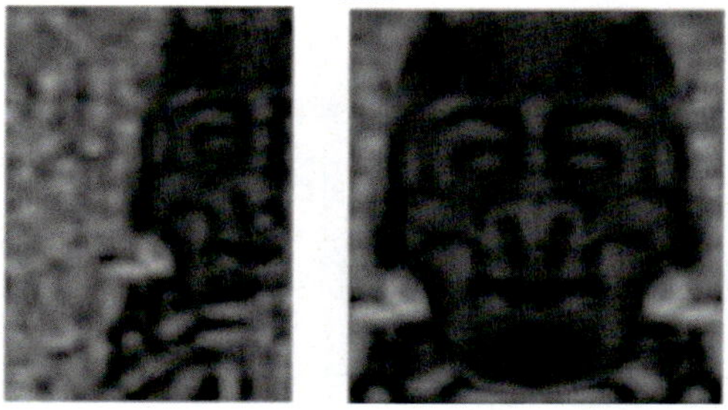

The next photograph shows two Bigfoot, one looking over the top of the other ones head. I do not know if the Bigfoot in the back is taller than the one in the front or if it could be climbing on the back of the one in front.

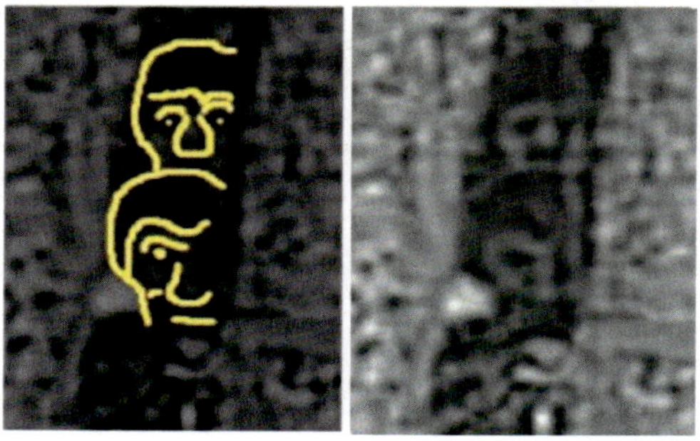

The Bigfoot in the back has a large bulbous nose and protruding brow. The Bigfoot in the front is more visible. It has the classic look of the Bigfoot from this area. Pronounced brow and the goggle affect around the deep set eyes.

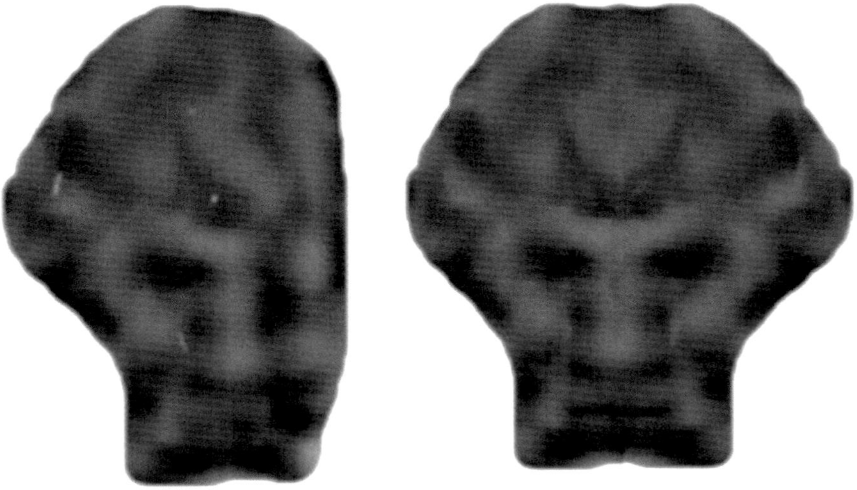

The above composite photograph is a reasonable facsimile of what the Bigfoot looks like

The final individual Bigfoot makes his way between the trees late in the video. It appears alone and is the most human looking.

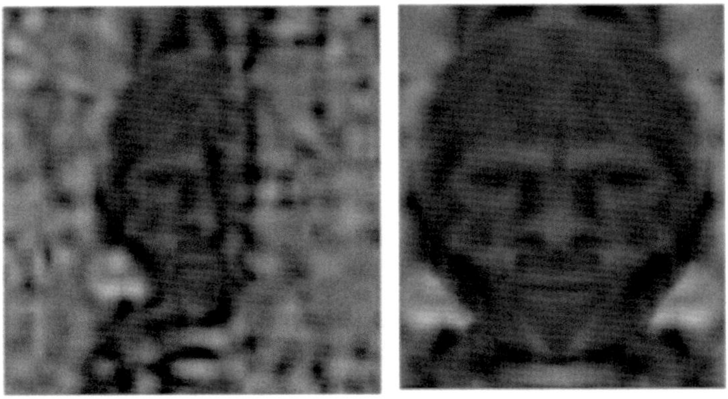

The above photograph on the left shows the Bigfoot peeking between the trees. The composite on the right is what the full face would look like. It has a pronounced brow and high cheek bones.

This is the last individual I can identify on the trail camera video. The trap worked, though not as I planned. The Bigfoot's curiosity is what allowed me to get the above pictures. They skillfully disabled my original trap but just had to see what those boxes on the trees were all about. Thankfully for me "curiosity killed the cat".

The below photographs are sketches of the Bigfoot "peeking" at the trail camera

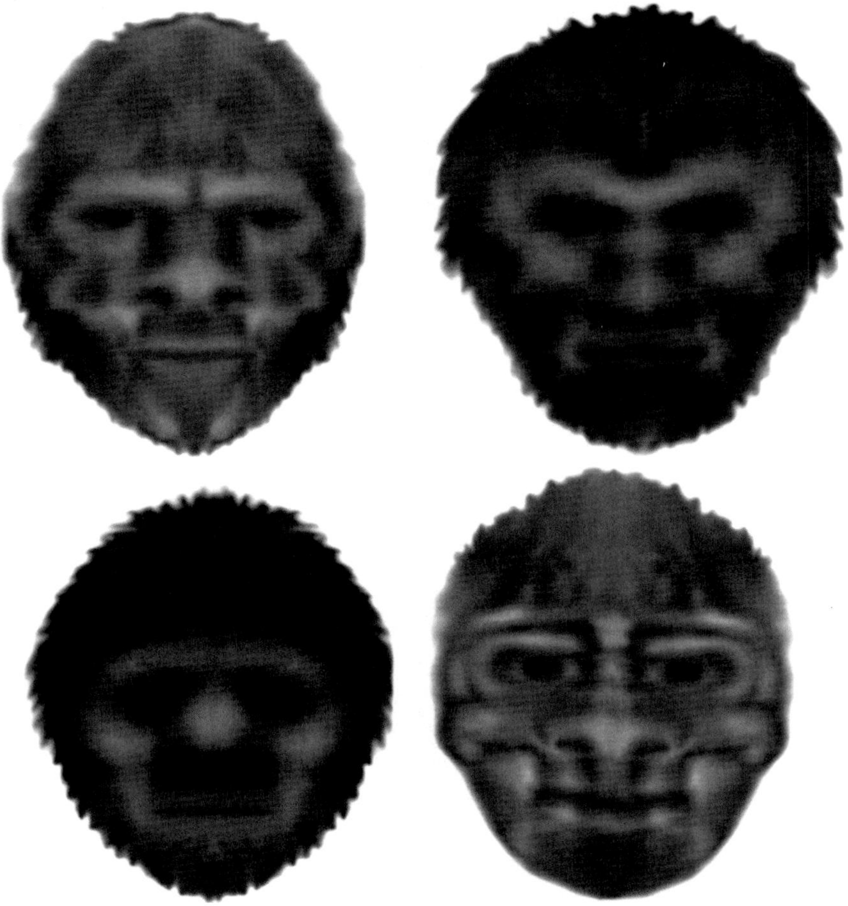

After we switched out the SD card and changed batteries on all the cameras we started looking around the corridor area and noticed additional snapped off saplings and a footprint in the middle of the small stream that runs down the middle of the corridor. It appears the Bigfoot were not happy with all of the trail cameras in this area.

After we documented the footprints and broken trees we check the feeding station. The rocks had been disturbed and the granola bar and apples were missing. We documented the feeding station and then we left the area without incidence. This accomplishment of capturing several Bigfoot on a trail camera was the culmination of over a year of research efforts. I had no one to guide me, no manual to follow, and no classes that I could have taken.
The entire process was trial and error, more common sense than anything else. Low tech methods worked the best in detecting their presence. In the

end the Bigfoot out smarted himself and saturating the area with every camera I had worked. Over the next few weeks the Bigfoot would continue to investigate the trail cameras. They would twist them, sniff them, sneak up on them, and even licked them.

Trail Camera Footage from the Trap

A Bigfoot rises up and places eye into the lens of the camera September 20th, 2010

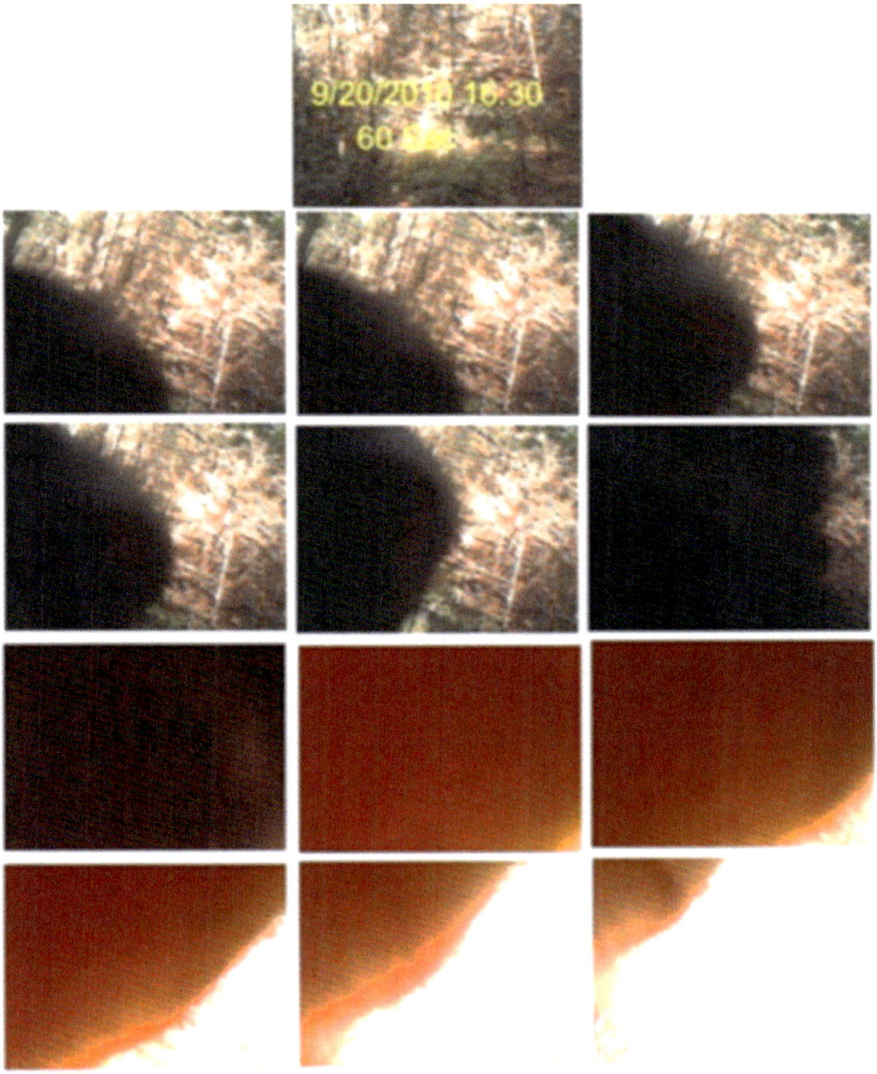

Nighttime silhouette October 5th, 2010

A Bigfoot investigates this camera thoroughly. Unfortunately it is so close to the camera all that was recorded is the infrared white out silhouette. It first sets the camera off and slowly sticks its head out from the cameras left side.

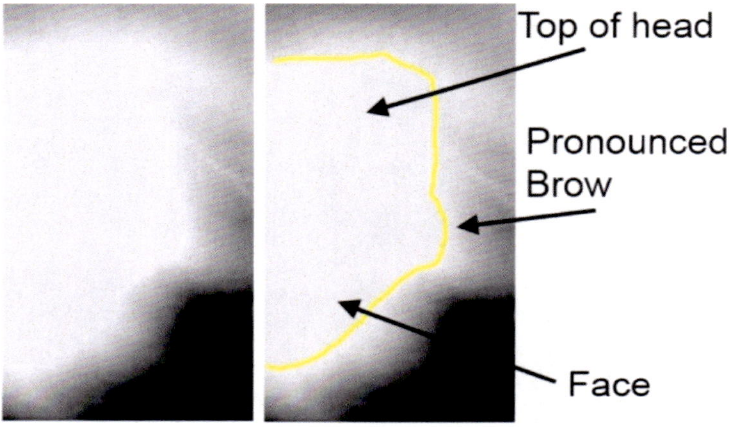

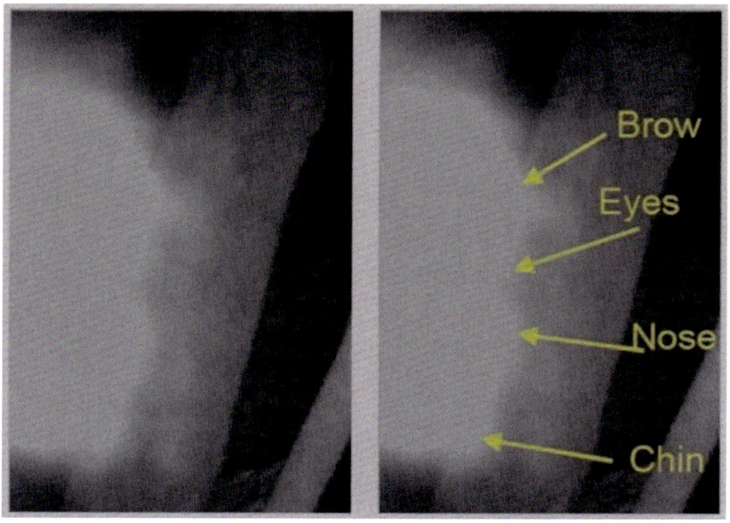

After investigating the camera from the side the Bigfoot then jumps out in front of the camera and makes erratic movements as if the camera may discharge something dangerous. The Bigfoot jumps out in front of the camera and lands with its left side facing toward the camera. After testing to see if the camera could be harmful the Bigfoot very slowly peeks out at the camera with just one eye from the left side. The corner of the left eye is visible in this still capture from the video.

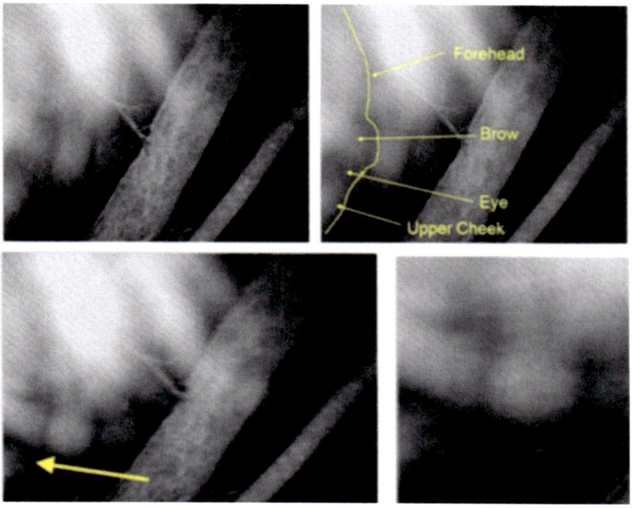

I also noticed what appeared to be a Bigfoot hiding between the two trees directly in front of the camera.

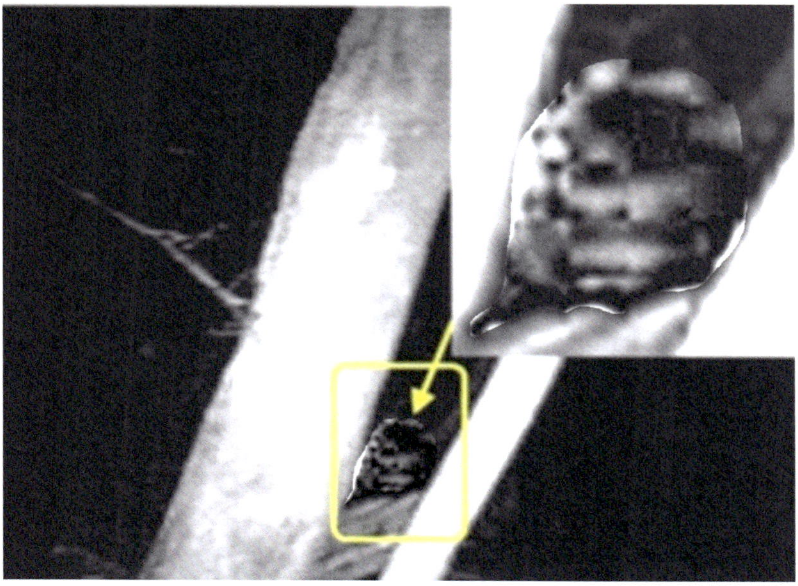

Bigfoot licking the Trail Camera September 20th, 2010

On this day a Bigfoot grabs the camera and shakes it violently and then it rises up and quickly licks the trail camera. The licking action is amazingly fast. If the Bigfoot would have moved slower I would have had a clear close up video of a Bigfoot's face.

The following is key frames from the video. The below sequence is the head rising up and the Bigfoot turning toward the camera. The Hair is hanging down in its face and the face is in the shadow of the tree but you can see the nose in one frame.

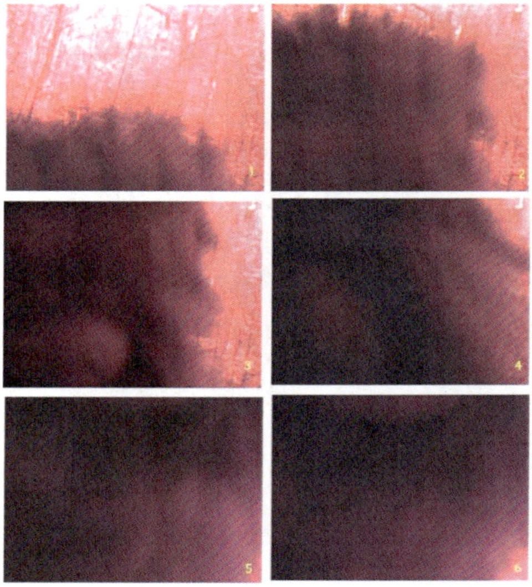

The next four frames show the chin of the Bigfoot as its moves in close to the camera

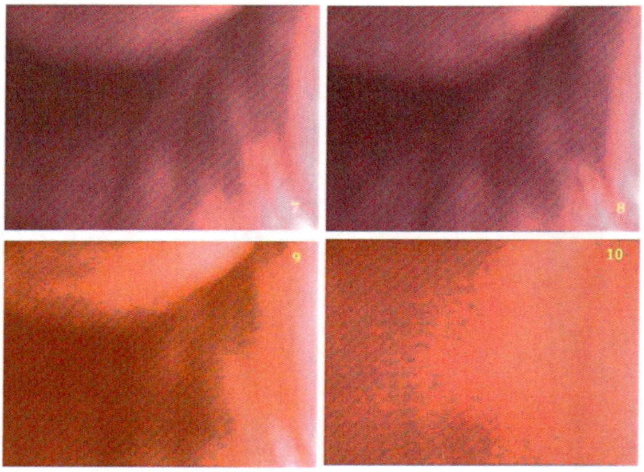

The next set of still captures is the actual licking sequence. You will see the tongue come out of its mouth and lick the lens.

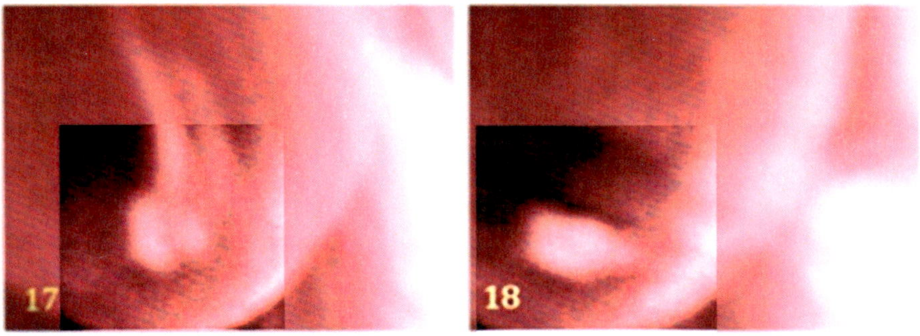

Enhancements of the licking sequence

Note how the face is distorted because of the speed at which the Bigfoot is moving. The entire lick sequence takes just over one second. Also note that the camera is encased in a steel cover. The edges around the lens and the IR emitters are very sharp. The camera was mounted between eight and nine feet on the side of a tree. The Bigfoot is looking straight into the camera!

Below are frames 18 and 19

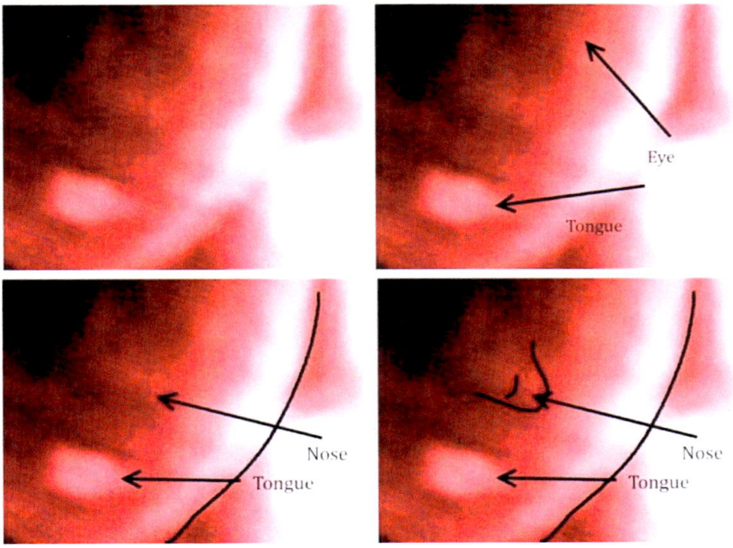

In the final set of still captures the Bigfoot moves to its right then in the last frame snaps its head to the side giving a good profile view of the eye, nose, mouth, and forehead.

Enhancements of the "Profile" and sketch based on still capture

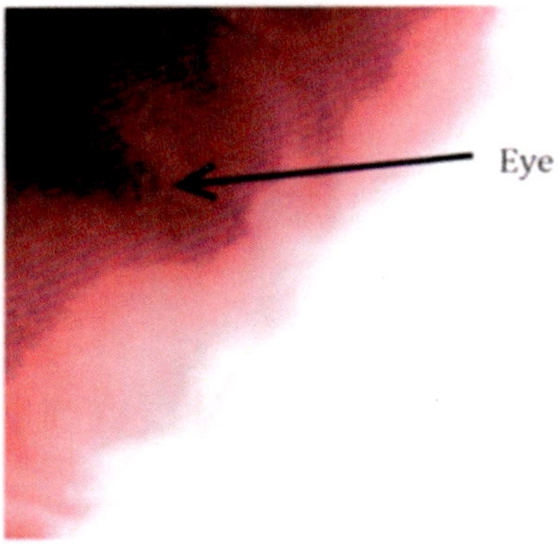

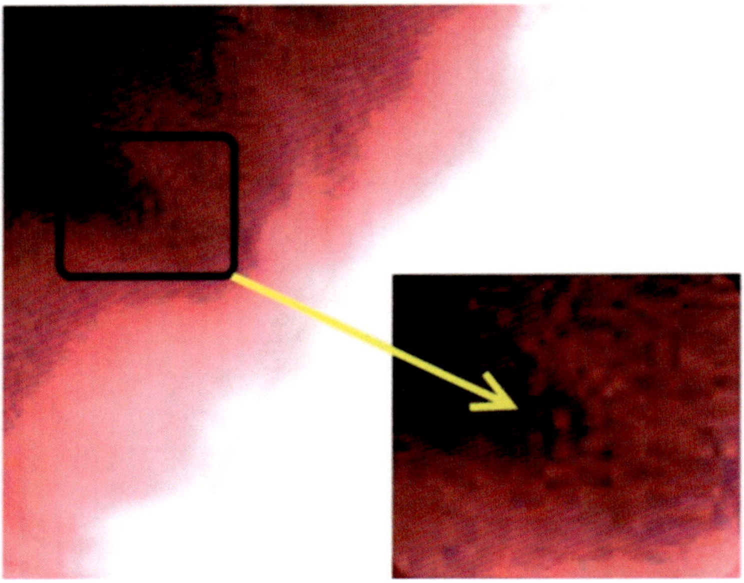

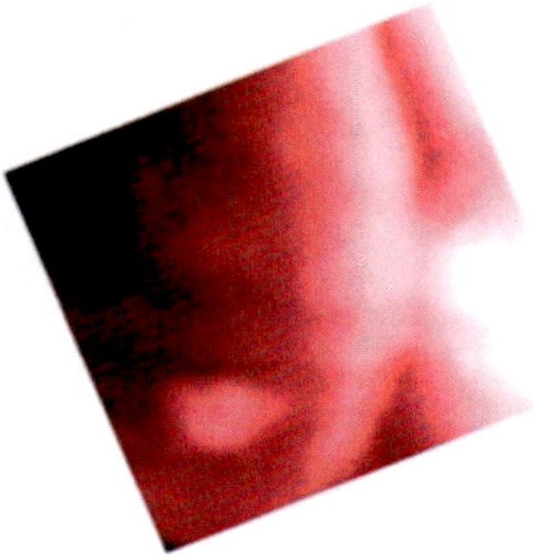

Chapter 42 - Unintended Consequences of Bigfoot Research – The Beginning

July 2010 – April 2011

After a year of investigating the Bigfoot in Research Area A I was becoming familiar with their habits and they were becoming used to me. They had become comfortable enough with my presence to exchange gifts with me, make vocalizations, allow me to hear them following me, and on one occasion one of them briefly allowed me to see it.

I was comfortable with this arrangement. I would visit them almost weekly and try to capture them on my "trusty" video camera or on one of the many trail cameras that I had placed in the area. This was a recreational activity for me much like fishing and hunting. I was soon to find out my "friends" had other ideas.

Starting around mid-July of 2010 I began to notice some "strange" occurrences around my home. It all started with Blue Jay feathers. I went out to my pick-up truck that was parked in the driveway to commute to work. I noticed beside the driver's side door on the ground was a single Blue Jay feather. We have bird feeders in our front yard to attract birds so I thought nothing of it. I picked the feather up, put it in my truck, and proceeded on to work. The next morning when I went to my pick-up truck again there was a single Blue Jay feather lying on the ground beside the driver's side door. Now I began to wonder, but this could be a coincidence so again I picked up the feather and proceeded on to work.

A couple of days had passed with me finding no feathers. I usually park in the same location out of habit, but my wife and I had changed parking positions. She parked on the left side of the driveway and I on the right. The next morning when I came out to get in my truck there on the ground beside the driver's side door was a Blue Jay feather! Once or twice was coincidence, but three times had me thinking. That evening when I arrived home I parked my truck back in the drive closer to the road in a totally different location than I would normally park. The next morning when I went out to get in my truck there again on the ground was a single Blue Jay feather beside the driver's side door. This would be the first sign that the Bigfoot had found my residence and were visiting on a regular basis. This leaving of the Blue Jay feather continued off and on for approximately a month then it abruptly stopped, but not the strange occurrences.

The below photograph shows the actual Blue Jay feathers left for me each morning

Starting in September of 2010 my wife began to complain that the dog food was disappearing from the rubber tub we keep it in. We have three outside dogs in our backyard; a large Golden Retriever/Labrador mix, a Chow, and an adopted Jack Russell Terrier. We keep the dog food on a covered wooden deck in a rubber storage tub with a lid on it. The lid snaps shut so it takes some effort to open and close. My wife started noticing that the dog food was not lasting as long as it normally did even though she was feeding the dogs the same amount each day on the same schedule. A fifty pound bag of dog food would normally last approximately ten days. It was now only lasting five to six days. Suspicious of what could be going on my wife one evening right before dark marked the level of the dog food inside the tub. The next morning when she went out to feed the dogs the level of the food was down approximately three or four inches! The lid was tightly secured when she went in for the evening and it was still tightly secured when she came out to feed the dogs the next morning. Now we both were wondering if the Bigfoot were helping themselves to a few handfuls of dog food during the night. We also wondered how they were getting past the dogs. The mysterious missing dog food continues on to this day though not as frequent as in the winter months.

I am the first one out of bed in the morning. A part of my morning routine is to get the inside dogs out of their crates and take them out for their morning constitutional. In addition to the before mentioned outside dogs we also own two miniature Chihuahuas, a female Pitbull, and a Beagle/Dachshund mix. I begin to notice a change in their behavior especially the very protective Pitbull named Jasmine or "J" as we like to call her. I also began to notice out of place bird calls early in the morning. There was something just not right about them because they were just a little off. It sounded like they were being made by someone and not by a bird. J would also act very suspicious and sniff around the corners of the house. She also became very "jumpy" and would bark at shadows and out of place trash cans which was very uncharacteristic for her. I would also get the now familiar feeling that I was being watched on a regular basis.

My Pitt Bull J would wake me up to take her out in the middle of the night if she could not hold it until morning. Starting in September 2010 she would occasionally refuse to come off the front porch. This is after she woke me to take her out! I would have to force her to come down the steps and go to the bathroom. She would stay right beside me and do her business then immediately return to the door wanting back into the house. This was totally out of character for J. I do recall a couple times she refused completely to come down the steps and whined for me at the front door. Now the Bigfoot visit more often and at various times during the night.

On one occasion in October of 2010 it was approximately seven o'clock in the evening. It was a warm evening, and the windows were open as I worked on the computer in the down stairs den when I got a whiff of very strong Bigfoot odor. I immediately went outside with a flashlight and began looking in my trees and my neighbors trees. I am sure they must have thought I had lost my mind but I knew a Bigfoot was out there. I never did find it. After a few minutes I stopped my search before I was questioned by one of my neighbors about my "strange" activity.

In November of 2010 my wife began hearing scratching noises at the front door. She would hear them usually after 10:00 pm in the evening while she was watching TV in the upstairs den of our split foyer. I have been called to the den many times since to check out the noise. We have no outside cats or any that live close to our house. We live in a small city so I think I would have noticed a raccoon or possum around our house by now. I always thought my wife was somewhat paranoid until I heard the scratching on the downstairs den window. Unfortunately every attempt to catch the perpetrator has been unsuccessful. I even tried sneaking out the back door and around the house while the noise was being made and found nothing at the front door.

These scratching occurrences get more interesting when my mother-in-law heard the scratching after our Christmas Eve get together. We had finished up eating and exchanging gifts and were sitting around drinking coffee and relaxing for a few minutes before my in-laws needed to leave. I had gone into the kitchen to get a refill on my coffee when my mother in law exclaimed "do you hear that scratching? Someone it's scratching at the front door and jiggling the handle." She then asked my wife to have me check it out. She was afraid someone was trying to break in with us there! I looked at my wife and she made an excuse that we had a "cat" doing that around the neighborhood. I went and checked and as usual there was nothing there.

The next day being Christmas Day we all turned in early. Like most families after getting up early and opening presents we planned to travel to my parents to open presents so we all needed the rest. It was a typical Christmas day; we spent the afternoon at my parents then returned home about dark. We were all tired so the kids settled in playing with their new gadgets, my wife put in a DVD movie in the upstairs den, and I retired to the downstairs den to watch some TV and play on the computer. About ten thirty my wife summoned me upstairs with a loud calling out of my name. I quickly ran up the steps to find her at the top of the steps very agitated. "There was something scratching at the door and jiggling the handle of the screen door, you need to check it out, it was so loud it startled me!" I turned and the heavy wooden front door was locked and bolted. I have my wife turn on the outside lights which probably alerted any Bigfoot outside if the loud yell of my wife did not already do it. I thought about not turning on the light but to be honest I was not prepared to open the front door and confront a seven foot Bigfoot. When I finally opened the door the cold air rushed in and gave me a chill. There was a light fluffy snow falling. It was not lying on the ground but it was falling at a moderate pace. I sniffed the air and looked around the bushes in the front of the house and around the cars. I did not smell or see anything. The concrete porch was very cold on my feet so I stepped back into the house and announced that all was clear. This seemed to satisfy my wife and we all agreed it had been a long day and was time to go to bed.

I was lying in bed thinking about the night's events when it hit me, the back deck! I did not check the back deck or the backyard. So I hurriedly retrieved my video camera just in case and headed for the back door. I did not want to be like so many others that surprise a Bigfoot and have no way to capture it on video. I go to the door that leads to the deck, prepare my camera, turn on the light, and open the door quickly. To my disappointment I did not surprise a Bigfoot standing in my backyard, but I did notice something amazing. I had mentioned earlier about a light snow falling. The wind had blown a dusting of the snow on the front third of our wooden deck. There in the fresh snow is a single seven inch footprint. I was barefoot myself forgetting to put anything on my feet so I stepped out to get a closer look. About three feet from the side rail of the deck was this "perfect" seven inch footprint. I looked closer and I found another impression in the old carpet we had on the back deck under the glass table. The impressions stayed on the carpet and went around the backside of the table to the dog food tub. Then they disappeared. I took video of all the footprints, but only the one footprint on the deck would show up on the camera in the low light. My feet were freezing so I went back into the house to put on my shoes and get a coat so I could do some further investigating.

Once I returned I checked the rail on both sides and the snow was not disturbed. I also looked into the yard, but with the light blowing snow there was a wide path around the house that had no snow. The ground was frozen and I could find no other footprints.

I have thought about this and have come up with a theory on what happen. I think at least one adult brought the infant to my house. After messing with the front door as a diversion they crept around the back of the house. The adult lifted the infant over the rail and set it down on the deck. The infant was too small to reach the carpet without taking one step on the deck in the snow and leaving a print. It then made its way around the glass table walking on the carpet which had very little snow on it. The infant opened the dog foot tub and ate some dog food. The parent then went around the other side of the deck and reached over the rail to retrieve the infant or the infant jumped the rail on its own since there was no snow on this side of the deck or railing. I am leaning toward the infant jumping over the rail and returning to the other side where the adult was waiting. I say this because the side of the deck where the dog food tub is located is at the back right edge of the house and can be seen from the road. I also have a light with a motion sensor on this side of the house. I think the adult stayed behind the house and let the infant jump down and come back around. This is just a theory but a plausible one.

Below are still photographs of the seven inch footprint found on the deck

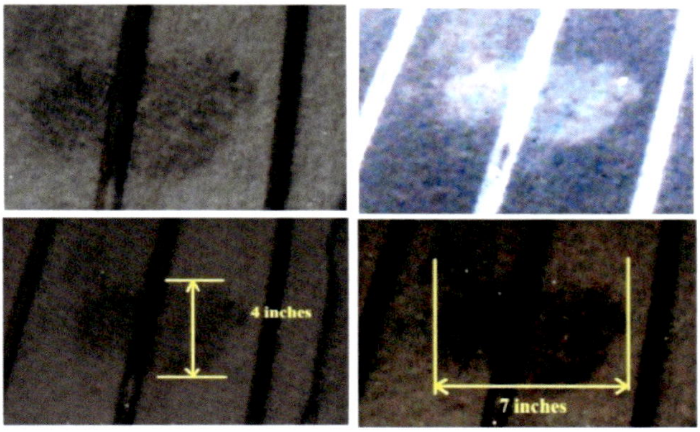

After this event I began a cat and mouse game with the Bigfoot. I had a couple older game trail cameras so I began to place them around the house trying to capture a picture or video of my regular visitors. To my dismay no matter what I did the dog food continued to disappear and all I could capture on the trail cameras was the white wash of something directly in front of the camera.

I decided to take a new approach. I knew that the Bigfoot were walking around the sides of my house to stay out of the lights and avoid leaving any footprints when it snowed. I have a fireplace and the byproduct of course is a fine gray ash. So on January 18th, 2011 I decided to sprinkle the ash around the corners of the house to see if I could catch a footprint of a Bigfoot. I awoke the next morning and went out before daylight to check the ash for footprints. To my shock and surprise it had worked!! In the ash on the front left corner of my house was a very clear footprint! I ran back in the house and retrieved my video camera and began to take video. The weather report was for rain to move in later that morning so I did not want to wait. I also could see where the ash had stuck to the bottom of the Bigfoot's foot and it had left residue in the shape of a foot in the grass as it headed towards the back of the house and over the four foot dog fence into the backyard. The track was 15 inches long. The ball of the foot was five and a half inches wide. The heel of the footprint was four inches wide.

The toe and heel impressions were visible. I documented the footprint and then went on to work. My wife went out after daylight and documented the footprint as well, but the print had already begun to deteriorate due to the light rain that was falling.

The below photographs are of the footprint I found in the ash beside my home

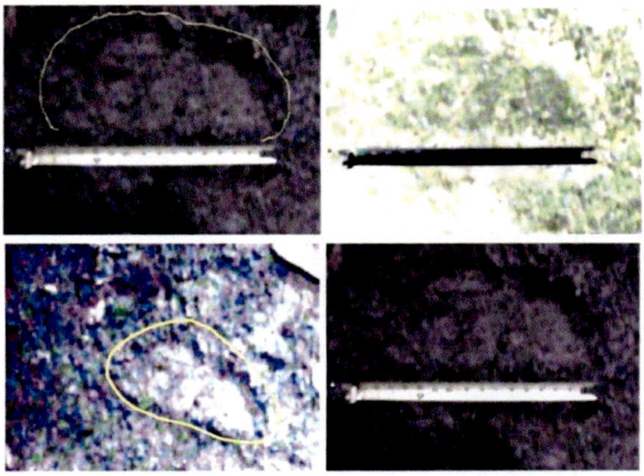

The below photographs show the "Big Toe" impression before the rain degraded the track

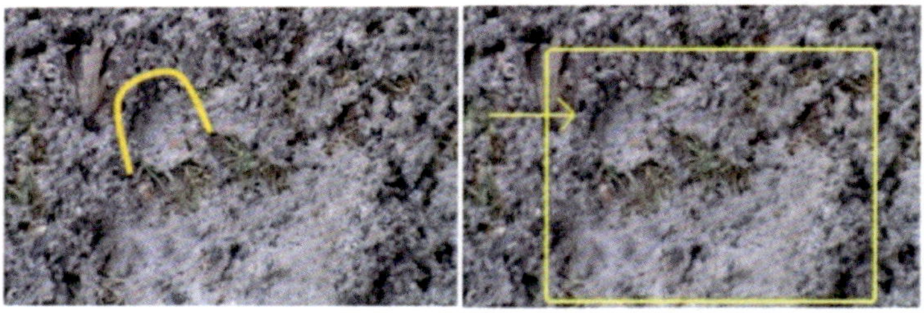

The below photograph is the footprint made in the grass by the ash residue that stuck to the bottom of the Bigfoot's foot.

I continued to put the ash out when the weather would cooperate but the Bigfoot were on to me and I have not been unable to get them to step in the ash again.

We would continue to allow the Bigfoot to eat our dog food as the winter wore on, but not because we wanted to feed them but in fear of what might happen if we removed the dog food from the back deck. In February of 2011 the Bigfoot began to develop a pattern of visiting approximately once every ten days. The scratching on the door stopped and the dog food was being taken less and less. In early March of 2011 my wife threw out about a dozen over ripe clementines into the dog lot where the motion sensor floodlight was located. When I ask her why she threw them out given the Bigfoot issue she said she was "not thinking" and was hoping the squirrels might eat them. Instead something very strange happened in the middle of the day on March 7th, 2011. The day was warm for early March and my wife had opened the windows. It was sometime around one o'clock in the afternoon when I get a phone call at work. It is my wife and she is very agitated and scared. I asked her what was wrong. She said she heard the dogs start barking loudly then all of a sudden they stopped. She got up to go check and when she stood up she could see from the front window that several of our outside dogs were running loose in the front yard. She rushed out front door and down the street to get the dogs. When she gathered them up and returned home she noticed that the threaded chain link we use to secure the latch on the gate had been unscrewed and thrown on the ground several feet away. She also noticed all the clementines she had thrown out were gone. She placed the dogs back into the backyard, retrieved the threaded link, and placed it back on the latch. I calmed her down and told her to shut the windows and call the police if she felt in danger because they could get there before I could. She refused to call the police and said she was more "flustered" than anything. She said they "fooled me and snuck right in there in the middle of the day!"

I can verify that the clementines where in the backyard less than 24 hours prior. The previous afternoon I had gone into the backyard to pick up a tree limb that had fallen and take it to the road. I had complained about the clementines that were not being touched by the squirrels and that in a few days they would begin to spoil and smell. I also remember shutting the gate and placing the threaded chain link all the way down. I remember because my wife made a point to remind me to do this because if I left this out of the latch our Golden Retriever mix would flip the latch up and let everyone out, so with my wife supervising from the deck I put the link on and screwed it down tight.

Did the Bigfoot visit the house during the middle of the day? Were they savvy enough to let the dogs out to distract my wife so they could grab the clementines and make off with them? We know it was not any children since they were all in school. Our dogs would have violently defended the backyard from a stranger even if they did open the gate.

We have had more than one incident where an exterminator has let himself in the back not realizing we had dogs and was backed out of the backyard with the gate open by the Chow and the Golden Retriever mix.

Later in March of 2011 I decide to straighten up the backyard on a Friday evening. It was a warm day and the woodpile at the back of our lot was looking messy. We had burnt a large amount of wood during the winter and the metal poles that I had placed under the wood to keep it off the ground were scattered around the backyard. So I re-stacked the remaining wood, neatly stacked all the metal poles against the back fence, placed a tarp over the remaining wood, and stuck an old bird house up under the tarp as well. I left three green barbed wire fence posts in the ground. I placed them at the end of the woodpile as a stop so the wood could be stacked about waist high. I figured I would wait until it rained to loosen the ground so it would be easier to remove them from the hard ground.

Saturday was a nice day and I spent it in the woods that morning then we had a cookout that afternoon when I returned home. My wife complimented me on how much better the backyard looked since I had cleaned it up. That night around three o'clock in the morning I was awaken by three loud raps on the window. I had fallen asleep on the couch in the downstairs den so these raps startled me. I woke up and proceeded to get my gun and then looked out the ground level window. I saw nothing so I proceeded to the front porch where again I saw nothing. I suspected it was the Bigfoot but what did they want? After I calmed down I went back to bed and there were no more incidents.

On Tuesday of the following week my wife alerted me that someone had been messing around in the backyard and had scattered the round fence poles and left a "mess" in the backyard. The weather had been very cold and rainy both Sunday and Monday. The children had not been outside either day. When I went outside to clean up the mess I found something very odd. One of the round metal chain link post appeared to have been laid out on the ground with one of the green barbed wire fence post laid across the lower third forming a cross. The green post had been pulled out of the ground. I also found the bird house had been pulled out from under the tarp. There were three sticks of almost equal length laying on the right of the round post under the green fence post. They were arranged in a W formation. The bird house was on the right side of the round pole above the green fence post. Directly below the round fence post about two feet away but directly aligned was a small stick about six inches long.

The below photograph shows the odd formation I found in the backyard

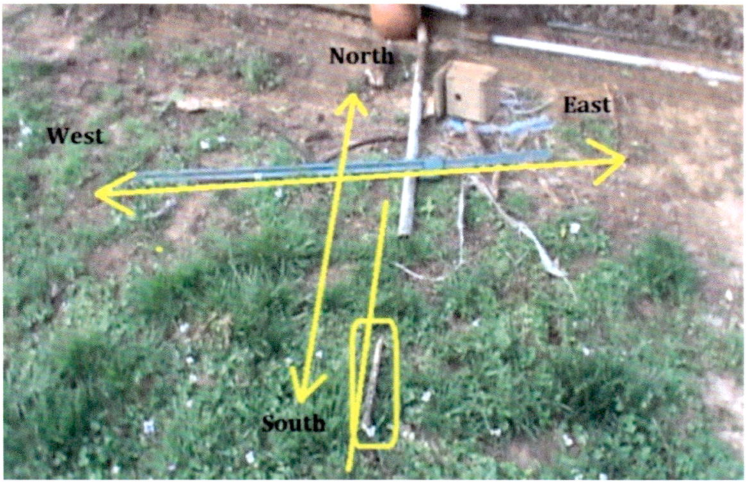

What is this? I thought to myself. It did not look like a random occurrence! Who pulled the fence post out of the ground? How did the bird house get out from under the tarp? Was this left by the Bigfoot? Does it mean something?

I thought on these questions the entire next day then it came to me, check it against a compass. Maybe it is something they used to find directions? So the very next evening I went out in the backyard with a compass in hand and to my amazement the round pole was only a degree or two off from magnetic north. The green post was exactly aligned east to west. I then wondered if they had used the North Star to line up the silver pole that was only a few degrees off. The next clear night I went out into the yard and as I had suspected the round pole was aligned with the North Star!

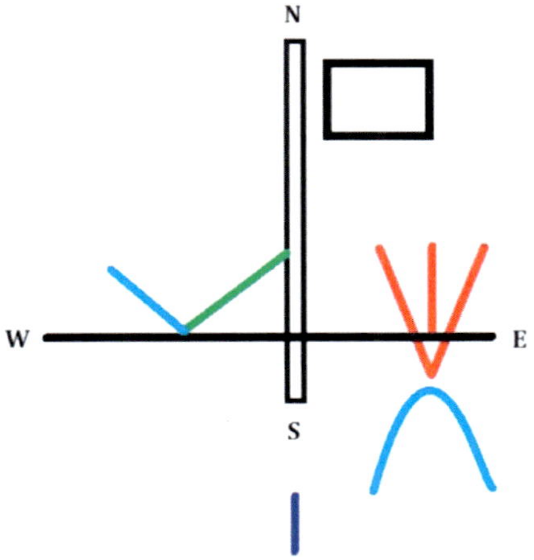

What does this mean? Did the Bigfoot lay this out on the ground to get their bearing? Is it some sort of message to other Bigfoot or to me? The red arrow in the below diagram is pointing south. Research area A is directly south of my home. Is the Bigfoot telling me this is where they are from? Many unanswered questions.

I begin to experiment with the trail cameras again determined to get a picture of the Bigfoot that were visiting my residence. I have one camera that can only take still photos so I set it up under the motion sensor light. My thinking is that the constant noise the flood light's motion sensor puts out will mask the sound of the trail camera's motion sensor. On March 30th, 2011 this strategy may have paid off. The camera is set to snap a photograph each time the motion sensor is triggered. One of the photographs taken on that night has a suspicious looking figure crouched beside a tree in the lot next door.

The below photograph shows a possible Bigfoot with enlarged and enhanced inserts

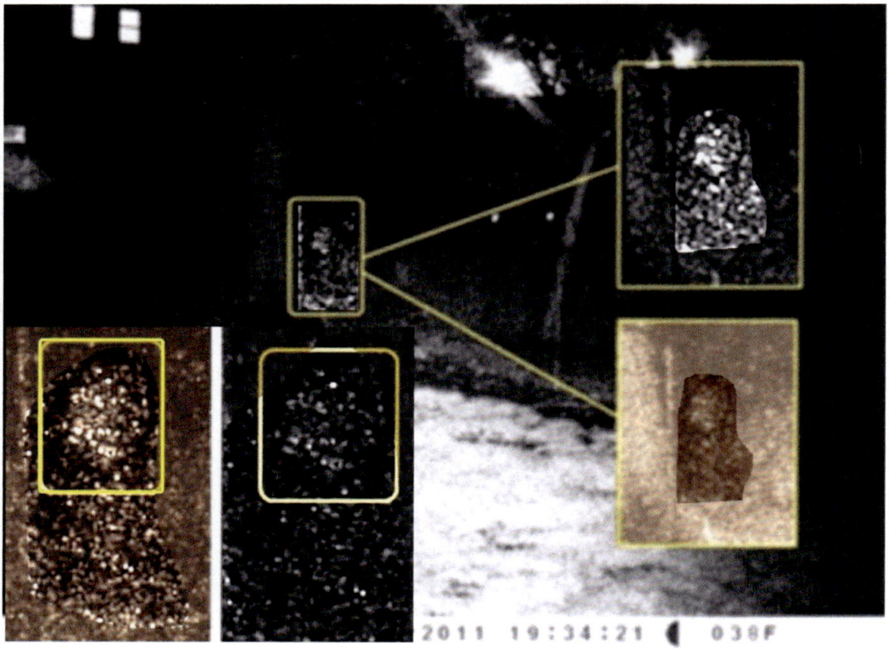

I thought that the possible face was caused by a clump of grass but when I align the nighttime photograph with a photograph taken during the daytime from the exact same position the grass clump is above the subjects head.

I wish the photograph was clearer but true to form the Bigfoot always manage to stay back out of the IR light.

The very next night something large stood directly in front of the trail camera. Being so close the IR light was reflected back into the lenses causing a complete wash out of the detail. The interesting thing about this picture is that the trail camera was sitting on the rail of the deck. In order to stand on the ground and look into the camera one would have to be at least seven feet tall.

The Bigfoot now know where I live and are visiting my home on a regular basis. I researched this phenomenon and discovered it is a common occurrence among Bigfoot researchers; the Bigfoot will follow you home!

May, 2011

I have been trying unsuccessfully to capture my visitors on a trail camera. No matter how hard I try or how I place the trail cameras I am only getting "white out" smears. My luck changed on May 4th, 2011 and I captured eye shine. I can identify four individual Bigfoot in the series of nine photographs taken by the trail camera.

The camera is tripped at 20:58:33 and begins taking a burst of nine pictures. In the first photograph three individuals are visible.

In the photograph above subject number 1 is at least eight feet tall. The entire window behind the subject should be visible, but it is partially blocked by its head. There is also a basement light that should be visible above the fence line and it is blocked as well.

A picture is snapped every nine seconds. The following cropped photographs follow the two Bigfoot's movement via their eye shine over the 80 second period. The timeline goes from left to right.

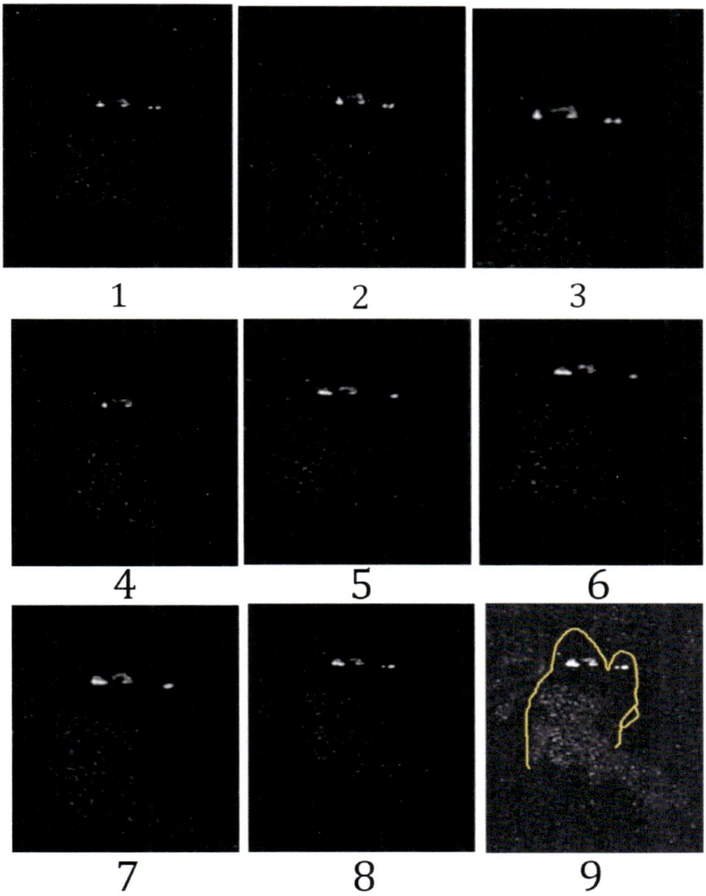

It is my opinion that the above sequence of cropped photographs shows the adult female of the group holding an infant on her left side. The female is rocking back and forth thus causing the blurring of the two eyes. The infant is sometimes facing the camera, and at other times looking away. This is why you will see two eyes, one eye, and then nothing. The teenage of the bunch appears for just one frame in the middle of the neighbor's backyard. Given there is only a single eye shine I make the assumption it is facing one direction or possibly has its back turned completely to the camera looking at the neighbor's home or garage area.

This was the only trail camera video I was able to capture during the three weeks in May. I wonder if the group of Bigfoot that is visiting my residence is the same group that followed us to our camping location back in October of 2010.

The day after the previous pictures were captured on the trail camera I was investigating the area in the corner where the large Bigfoot was hiding between the two trees. I found something very odd. A small limb had been twisted around a large limb apparently in an attempt to get it out of the way of its face.

The below photograph is a possible fourth Bigfoot standing beside the female.

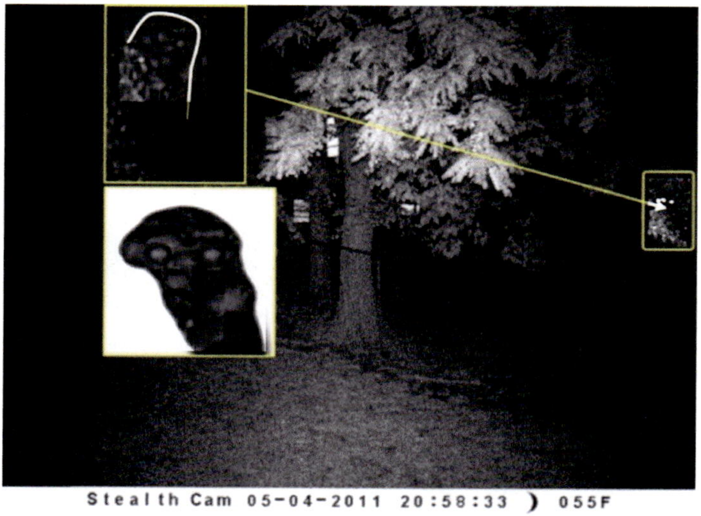

Gifts and Rocks

I had a conversation with a friend shortly after the above pictures were captured. He had been looking at the map and suggested I try to surprise the Bigfoot by putting some dog food and something familiar at the nearby river the Bigfoot follow out of the National Park. I suspected the Bigfoot were using the river to make their way out of the National Park and then on to my residence. I decided to take some of my wife's hair from her hair brush and tape this to a small plastic bag of dog food. The Bigfoot has shown an interest in my wife since her sighting back on Easter weekend. I took the bag of dog food and hair and placed it on a game trail beside the river approximately three miles from my residence.

The below photograph is the "package" I delivered to the Bigfoot

I took the hair and placed it at the river on Friday, May 6th. Approximately five days after leaving the "message" for the Bigfoot, think we received an answer. We mow our yard on Mondays and when we mowed our yard on Monday, May 9th there was nothing out of the ordinary. On Wednesday we noticed something out of place in the side yard beside the chimney. A flat, oval, river rock had been placed against the house beside the chimney. We could tell it had been placed there after we had mowed because there was fresh cut grass on the back of the rock.

Is this rock the answer to the "invitation" we had placed at the river? I thought it interesting that the rock was placed in the side yard instead of the backyard. But then I remembered I have had the game camera setup on the deck monitoring the backyard for the last two weeks. The game camera was being counterproductive so I decided to remove it and place Digital Voice Recorders (DVR) on the back deck. Other researchers have had success with them and I am continuing to lose dog food at a constant rate.

Hand prints on the siding

We had a pair of Robins nesting under the right front eve of our house. They were very active and had a nest full of eggs. On or about the 14th or 15th of May one Bigfoot stood on another's shoulders and took the eggs out of the Robin's nest. Our house needs painting the maroon paint has oxidized. If you touch the siding the paint will come off on your hands and leave a print.

The side of the house under the bird nest has multiple prints to include a large hand print, shoulder print, arm prints, and general scuff marks where the Bigfoot was leaning against the side of the house.

The hand print and the shoulder marks were ten feet high off the ground. The hand print is twice the size of a normal man's hand. The detail in two of the smudges is amazing. The oxidation makes an ideal medium to leave an impression in. The shoulder print has hair detail! The individual hairs can actually be identified.

The below photograph is a view of the side of the house to include where the many prints and scuff marks are located on the siding. The bird's nest is on the gutter in the upper left corner.

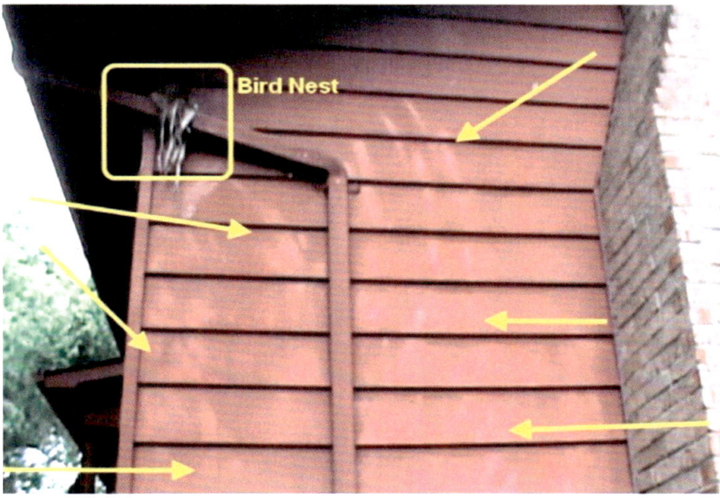

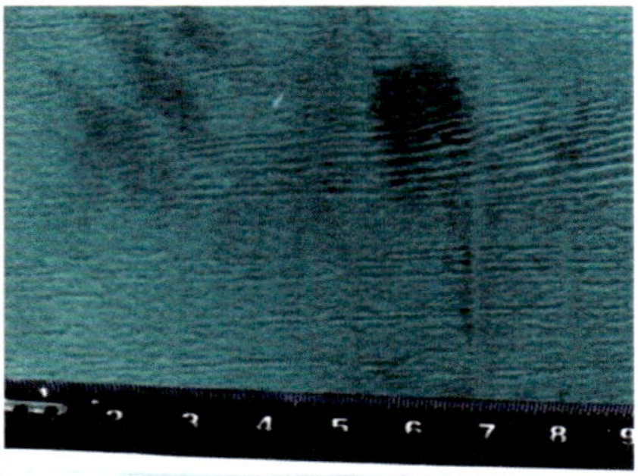

The below photograph shows the shoulder print in negative where the hair detail is visible

The Dog Food

The Bigfoot continues to regularly take dog food from the latched plastic storage tub. On average they will take approximately two inches off the top. I placed the digital voice recorder on the deck and on several occasions you can hear the Bigfoot open the tub then snap it shut.

The photograph below shows the container in the morning after the Bigfoot has visited. The dog food level the night before was level with the top edge of the container. The Bigfoot took about two inches of dog food.

Fingerprints

I washed the top of the container to remove all dirt and leave the top free of fingerprints. That same morning I found the dog food missing I also found three fingerprints and a palm print on the lid of the container.

The below photographs show the three fingerprints, both the actual and in the negative. Please note the large print on the right is at least twice as large as my thumb. The dog food is packaged with "food oil" to keep it moist. These fingerprints were oily and shiny, probably made by the Bigfoot while closing the lid after eating the dog food and getting the oil on its fingers. The below photographs are the finger prints and the palm print in the center of the lid.

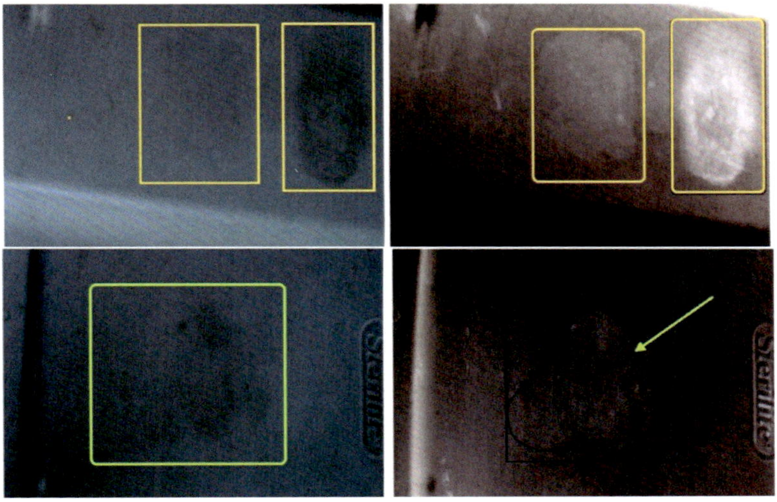

Bigfoot Hair

I placed a "hair trap" on top of the metal fence where the Bigfoot step over and come into my backyard. It has yielded one likely Bigfoot hair.

Dog Treat

We also placed a few dog treats on the patio table. The Bigfoot appeared to eat one dog bone biscuit and took a large bite out of the doggie jerky.

Conclusion:

Bigfoot continues to regularly visit my residence. They are taking dog foot and making some interesting noises as they move around on my deck. We kept hearing them make a "metal clanking" sound on the digital voice recordings for several days and could not figure out what the Bigfoot were doing. Then I was grilling hamburgers and when I hit the grate with the spatula my wife pointed out "that's the sound they are making." We have come to the conclusion that the Bigfoot were licking the grease off the grill surface and it makes a clanking noise as the grate vibrates!

I continue to gather evidence and try different strategies to get video of the visiting Bigfoot. I tried hiding a game camera in the dog food, but that did not work. On the digital voice recordings you could hear them unsnap the dog food container lid the immediately snap it back when they saw the camera. The motion PIR sensor will not trip with just the lifting of the lid, it is looking for a IR reflection, so lifting the lid will not trip the camera.

Two different neighbors report "Prowler in a black hooded sweatshirt"

May 2011

The neighbors that live directly behind my home are enjoying a warm evening and sitting outside after dark. One of the neighbors notices a dark figure in a tree adjacent to my backyard. One of the group, who is employed as a local sheriff's deputy, retrieves his flash light, walks up to the tree, and demands that the "subject" come down out of the tree. All four of my neighbors witness what they described as a "teenager in a black hoodie" jump out of the tree from over twelve feet high and hit the ground running. They were all amazed that it could jump from that height and not be injured. They were also surprised at how fast "it" ran, its speed was "amazing".

July 2011

The neighbor that lives behind me to the left heard their dogs barking loudly and stepped outside on their deck to see what was causing the dogs to bark. When she turned on the light a figure she described as a "tall, slender, teenager, in a black hoodie" came out of the shadows to her left and ran in front of her across the backyard in full view. She told me "it ran across the yard from left to right, and then it jumped the fence in one step and ran out of sight". The fence is four feet tall. When we first started talking about the incidence she claimed that we have a "Peeping Tom" in the neighborhood, but when she described the sighting she call the Peeping Tom "it" twice. She did call the police and reported a prowler. The police came to my residence to search the area since it was reported to run near my house. She also told me that a family that lives a few blocks away has seen this same "Peeping Tom" looking in the window at their sixteen year old daughter.

Analysis

- Two different groups of people give an identical description of a "prowler".
- The subject leaps twelve feet out of a tree without injury, runs off at "amazing speed", leaps or steps over a four foot fence in one step, and runs away at a "fast speed".
- One witness refers to the "prowler" as "it".
- The events happen two months apart.

I have come to the conclusion that a family group is visiting my residence. I think this group consists of a Male, Female, Teenager, and Infant. The teenager is the one that has been seen twice. He is approximately six feet tall, weighs 200 lbs., and is black in color. I was privileged to see him October 2011 for a few moments as he ran across the neighbor's backyard early in the morning. He went from a squatting position to a full run in moments and covered over 100 feet in mere seconds. He looked like a black blur.

Video Camera Footage

The Bigfoot regularly defeat my efforts to capture them using traditional trail cameras. I executed a plan to place a web camera on my deck. Then at random times during the night I would turn the camera along with the lights around the deck. I was hoping to capture a Bigfoot sneaking up on the deck to take dog foot. I would get up two or three times during the night to do this. I was able to maintain this schedule for two months, May and June of 2011. Most of my efforts resulted in fleeting shadows but I was able to capture an interesting clip during one evening in May of 2011.

The below are still captures of the deck and the area where the subject is located with enhancement inserts.

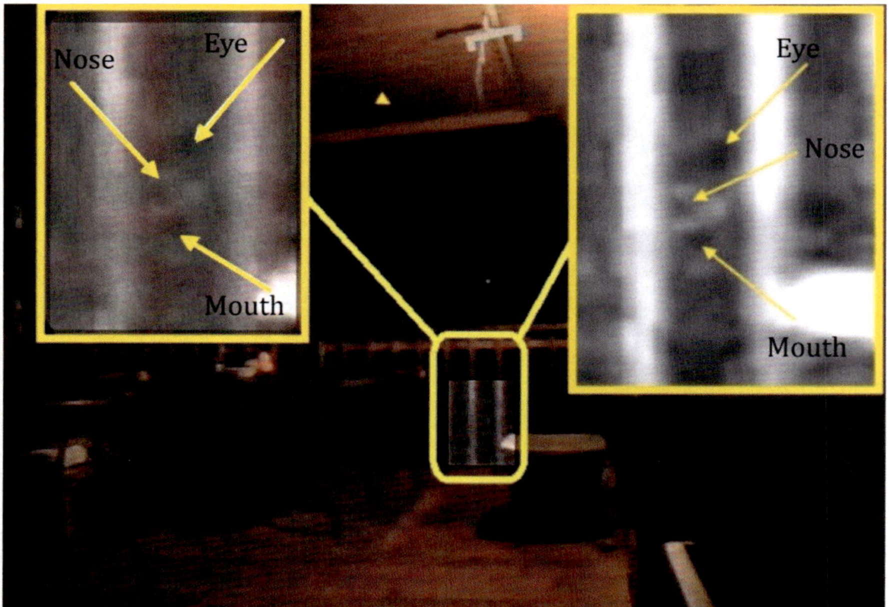

Footprints

The Bigfoot left two different sets of footprints in my backyard. Both footprints were found in the same area beside the deck. The first footprint was found on September 2nd, 2011. The print was huge! The width of the foot across the toes was eleven inches!

September 2011 Footprint

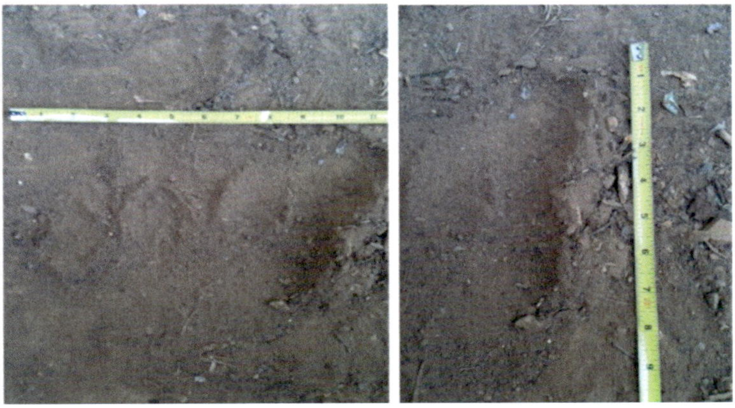

The big toe is almost seven inches long

October Footprint

The footprint below was left beside my deck on October 17th, 2011 shortly after dark. The footprint was fifteen inches long.

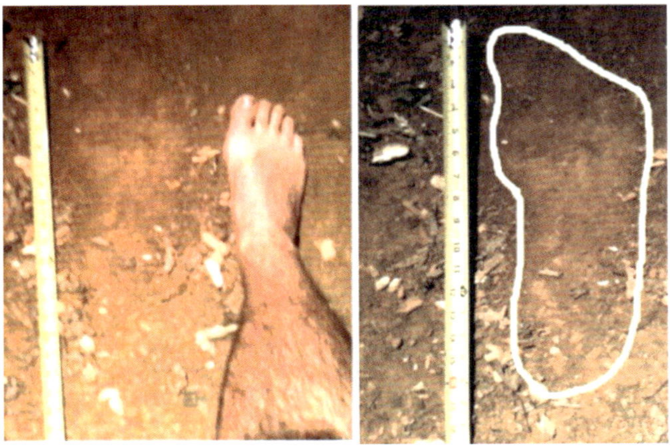

The Bigfoot continues to make visits to my residence. During the summer the frequency of the visits diminish to approximately once a week.

DNA Study Results Validate Residential Visitations

Three hair samples collected from my residence are in the Ketchum DNA Study. The samples are listed in Supplemental Table 1. Samples 2, 3, and 70 were verified to be Bigfoot/Sasquatch.

Below is an extract from the actual table from the DNA Study of only my samples.

Table 1. List of Purported Sasquatch DNA Specimens

#	Location	Circumstance	Submitted	Sample #, Type	Submitter Notes	Original Finder
2	Tennessee	Hair trap	July 2011	2. Brown Hair	North America Bigfoot Search (Dave Paulides)	Scott Carpenter
3	Tennessee	Hair trap	July 2011	3. Hair	North America Bigfoot Search (Dave Paulides)	Scott Carpenter
8	Tennessee	Hair in hair trap	August 2009	11. Brown Hair	North America Bigfoot Search (Dave Paulides)	Scott Carpenter
29	Smokey Mountains Tennessee	Tape trap on tree trunk	October 2010	38. Large amounts of short curly dark brown hair	North America Bigfoot Search (Dave Paulides) Biped observed watching campers from behind a tree on numerous occasions. Tape was placed on tree at 8 feet and hair was retrieved after it stuck to the tape.	Scott Carpenter
32	Eastern Tennessee Smokey Mountains	Tape trap on limb 6' above the ground in sighting area	February 2011	42. Reddish brown to brown hair	North America Bigfoot Search (Dave Paulides)	Scott Carpenter
43	Tennessee	Tape trap on limb 6'	April 2011	70. Black Hair	North America Bigfoot Search (Dave Paulides)	Scott Carpenter
50	Tennessee Smokies	Hair trap on a tree	May 2011	82. Light brown to blonde	North America Bigfoot Search (Dave Paulides)	Scott Carpenter
94	Tennessee Smokies	Hair trap on a tree	May 2011	122. Mixed color	North America Bigfoot Search (Dave Paulides)	Scott Carpenter
95	London Co.	Hair trap	July 2011	125. Hair	North America Bigfoot Search (Dave Paulides)	Scott Carpenter
96	London Co. Tennessee	Hair trap	July 2011	126. Hair	North America Bigfoot Search (Dave Paulides)	Scott Carpenter
97	Tennessee	Hair trap	July 2011	127. Hair	North America Bigfoot Search (Dave Paulides)	Scott Carpenter

This is one of the hair samples confirmed to be from a Bigfoot. It was collected at my residence.

Chapter 45 – Theories - The Eyes Do Not Lie – Bigfoot has the ability to see light in the Infrared Spectrum

Background:

Like many Bigfoot researchers I have attempted to capture video of a Bigfoot using trail cameras specifically designed for deer hunters. The units are equipped with a camera that can take still shots or video. The camera can capture images either in the night or during the day. The cameras use infrared (IR) light emitters and an infrared filter for the camera lens to capture photographs and videos after dark. The batteries will last anywhere from two weeks to over a year depending on the model. At first glance this looks like a great tool for obtaining video or photographic "evidence" that Bigfoot exist. The trail camera can be in the woods monitoring a likely area twenty four hours a day seven days a week. If the Bigfoot walked in front of the trail camera at night the IR emitters would illuminate the area with IR light that cannot been seen with the human eye. If camouflaged correctly the Bigfoot would never know the cameras were spying on them, or so I thought.

I soon learned why no one to date has captured a clear photograph or video of a Bigfoot using a trail camera. The Bigfoot can locate them with ease and avoid them. In my case they found them, investigated them, learned they would not harm them, and then came up with ways to circumvent them. The Bigfoot would stand to the side of the trail camera and shake it or wave their hand in front of it. I even suspect they threw a possum in front of it to see what would happen on one occasion. Once they realized it was not going to harm them they begin to block the front of the trail camera. I began to get video clips of nothing but bright white infrared wash out. This can only happen if an object is placed a few inches in front of the IR emitters and the IR is reflected off the object and "blinds" the camera. The passive infrared motion detector (PIR) would only trip the camera one time if the object remains stationary in front of the camera due to its design. This would allow the Bigfoot to neutralize the camera. They could just place their hand in front of it allowing the other Bigfoot to go about taking food or messing with the objects I had placed as bait. Then when finished they would just remove their hand. This would result in a second video clip that would show nothing but the area void of any Bigfoot. On many occasions I had this circumstance happen with my camera.

The only good information I could take from this behavior was it gave me the length of the stay in the area because I would get the "washout" when they arrived and the uneventful video when they left. This allowed me to estimate how long they stayed at the location.

I had painstakingly camouflaged my trail cameras. Placed them in locked metal cases with foam insulation to try and minimize any sound they might make. This left me asking the following questions: How did the Bigfoot locate the trail cameras and how did they know that the lens was in the front?

Theory: The Bigfoot can see light in the infrared spectrum and can hear the electronic emissions that trail cameras produce.

The first statement of my theory is more important in its ramifications than the later statement. If the Bigfoot can see infrared light then almost any attempt to mount a nocturnal expedition will be useless if the investigator is using any type of infrared light source. The Bigfoot would see the investigator as if he/she were shining a very bright flashlight into the darkness. This presents investigators with a major issue because most of the technology currently available for nocturnal use relies on an infrared light source.

The question now becomes what brought me to this conclusion that the Bigfoot can see infrared light other than the anecdotal evidence given in my background statement.

In Research Area A I had placed trail cameras at the "stick shelter" location. I have had sightings, footprints, and gift exchanges at this location with the Bigfoot. The trail camera was camouflaged and placed in heavy brush approximately three feet off the ground. The camera was aimed into the "stick shelter" where I had placed rock arrangements and food as an attractant.

In the early morning hours of October 24th, 2009 the trail camera was shaken, this tripped the PIR sensor and the 3.0 MP Wildview Trail Camera began recording. A couple seconds into the video clip a large hairy arm and hand are waived very quickly in front of the trail camera. The motion was so fast that it only lasted two frames. The camera records video at thirty frames per second so this was a very fast motion.

When the hand/arm passes in front of the IR emitters it causes a very bright infrared flash. This flash is reflected back off the surface of the camera and then forward toward the "stick shelter" and the surrounding area.

There is a Bigfoot hiding in the back of the "stick shelter" in heavy vegetation. The Bigfoot appears to be leaning forward and resting its chin on a large limb and observing the other Bigfoot as it's investigates the trail camera.

After the IR flash subsides the face of the Bigfoot reflects brightly in the IR light while the hair on its body appears to absorb the light. The Bigfoot is pushing its face through small limbs and leaves in the tree and only a partial left shoulder and left hand holding on to the tree limb is visible.

During the video the Bigfoot moves its head slowly from right to left then back to the center directly facing the trail camera. The movement is visible and measurable. The eyes, nose, mouth, chin, and forehead all move in unison ruling out the possibility this is an optical illusion caused by shadows and leaves being blown by the wind.

While reviewing the video one frame at a time I noticed that in some of the frames the eyes of the Bigfoot were visible. During the bright flash when the IR light reflected off the camera I noticed that the Bigfoot's left eye's pupil contracted to a pinpoint.

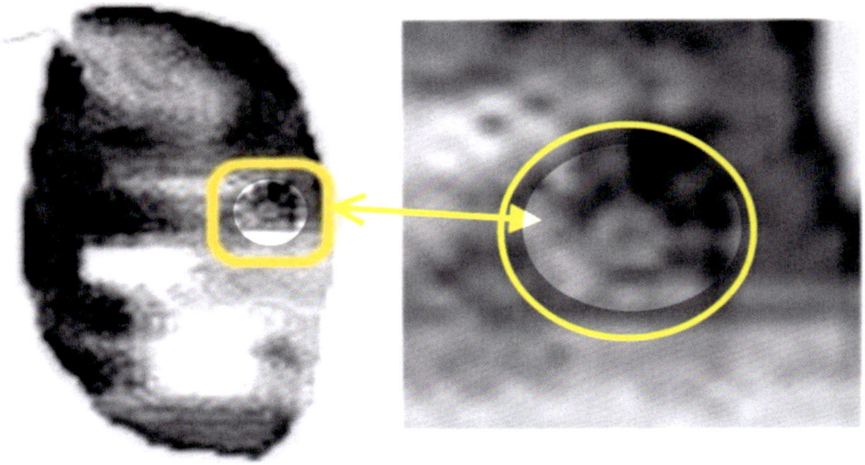

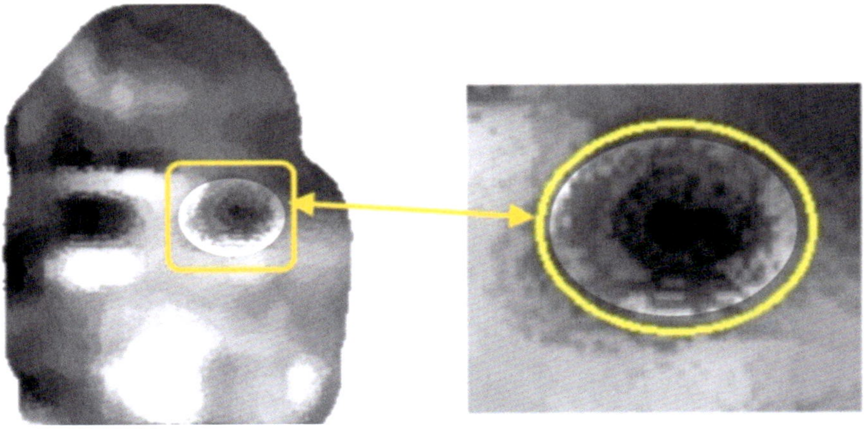

After the IR flash I continue to watch the eye of the Bigfoot in the frames where it is visible. Approximately fifteen seconds later the pupil has dilated to a much larger size involuntarily adjusting to the change in the IR light level.

Conclusion:

The Bigfoot's eyes adjusted to the flash of the IR light. I can only conclude that the Bigfoot has the ability to see light in the infrared spectrum.

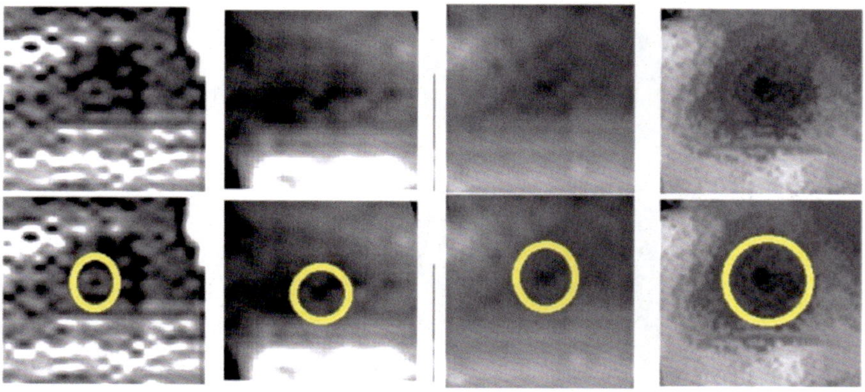

The above composite shows the pupil going from a pinpoint on the far left to being dilated in fifteen seconds on the far right. The ability to see light in the infrared spectrum gives the Bigfoot excellent night vision. All objects, trees, rocks, grass, animals and yes humans emit infrared light. This would mean that the Bigfoot could see in total darkness just like it was in the middle of the day. The photograph on the following page was taken with a special filter. This filter only allows infrared light through to the camera lens. This could be how the Bigfoot views the world after dark.

The ramifications of the Bigfoot being able to see light in the infrared spectrum are substantial. Researchers need to be aware of this ability and realize how difficult it will be to approach a Bigfoot at night without it being aware of your presence. The more I learn about the Bigfoot's abilities the more I am amazed!

Infrasound analysis of the "Sneering Bigfoot Video Footage"
Location: Research Area A

Background:

On April 30th of 2010 I captured a few seconds of a Bigfoot peering through vegetation and sneering at me. An interesting aspect to the video is that I did not discover I had captured the sneering Bigfoot until ten months later. I was reviewing the videos I had taken during the expedition that day when I found the video clip with the Bigfoot sneering at me. I have no recollection of the Bigfoot or the making of this video. I had what is known as a "missing time" experience. If not for the video camera I would never have known about the experience. Another odd aspect is even though I zoom in on the Bigfoot I never utter a word. I do not comment or react. I zoom in and out on the Bigfoot twice and then inexplicably turn and walk away!

I have a theory that I was bombarded by infrasound that caused me to "loose time" and then not remember what I had seen or done. In order to test this theory I needed to find software that was designed to detect infrasound. The human ear cannot detect sounds below 20Hz but we are sensitive to the effects of these sounds or better vibrations at or below 20Hz. I fortunately was able to locate freeware called "Audacity", which is enhancement software for music. It specifically was designed to allow you to filter and "clean up" your recordings and to improve the sound quality. One of the functions of the software is to detect ultra-low frequencies (infrasound) and allow them to be filtered out because this will cause distortion in your recording. This function would allow me to detect if infrasound was present during the time I was capturing the video of the "Sneering Bigfoot". The software algorithm would not show a continuous output of the frequencies in real time but it would calculate the average frequencies during the video snippet selected. This will work for my purposes since I am interested in the presence of infrasound and the frequency/amplitude at which it is being created.

In order to measure the infrasound in real time I used a freeware program called "Wavepad". This software shows the frequencies and their amplitudes in real time. This would give me a more accurate measurement of the amplitude spikes during the snippets.

I also had to make sure my video camera's microphone was sensitive enough to capture sounds at or below 20Hz. The Sony HandyCam DCR-SR68 has a high quality built in Zoom Microphone. This is able to record sounds lower than 20Hz.

Now that I had my software and was confident that my hardware was able record infrasound, I started my analysis of the video for any possible infrasound.

Analysis:

I have broken the video down into segments based on how close I was to the subject before, during, and after the event. The filter's chart shows both the frequencies and the amplitude. The amplitude is measured in decibels (dB) with 0 being the level of distortion. This means the filter scale is based on negative dBs so the higher the negative number the lower the amplitude. Normal speech volume on this filter scale would be between -70dB and -50dB with loud rock music around -10db.

I used the video editing freeware "Avidmux" to extract the audio from each clip into a .wav file. The .wav file was then loaded into the filtering software and the analysis option was chosen. The filter levels were set to the following specifications in order to detect frequencies lower thank 20Hz:

Algorithm – Spectrum
Function – Hanning Window
Size – 16384
Axis – Log Frequency

The first snippet is a control sample. I extracted the audio file from video that was taken when I first arrived at the area and was entering the woods. I am approximately one half of a mile from the video location.

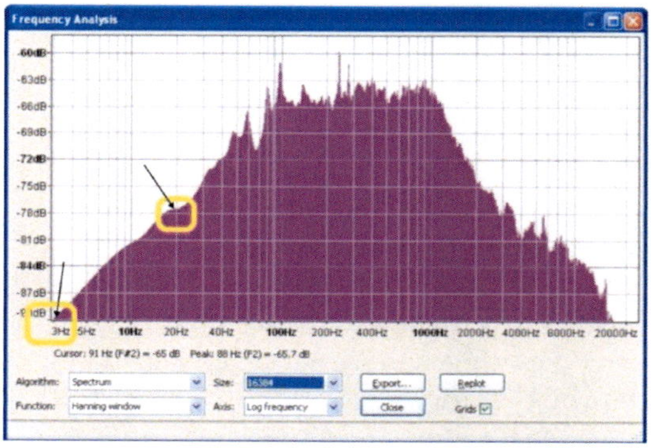

The infrasound amplitude average for 20Hz is -78db while the average for 3Hz is less than -90bB.

The second snippet was taken when I was still one hundred and fifty feet from the island where the Bigfoot was located and the event occurred.

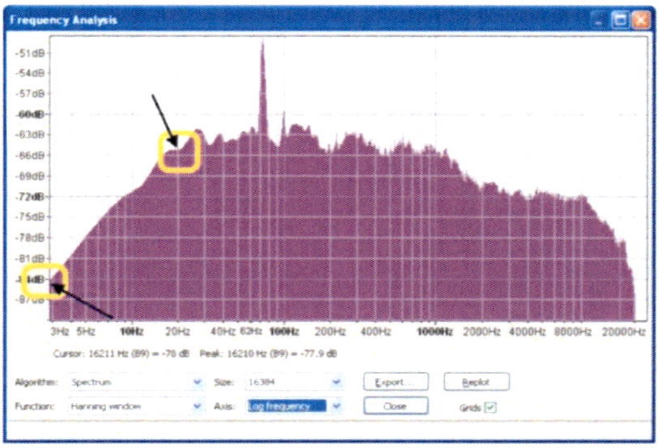

The infrasound amplitude averages have risen slightly. 20Hz is now averaging -65dB while 3Hz average has increased to -83dB. This is an increase of 12 dBs for 20Hz and 7 dBs for 3Hz. I am at the "Stick Shelter" during this snippet. I do not know if I may have had a Bigfoot around the "Stick Shelter" that was using infrasound to communicate with other Bigfoot or if it was trying to influence or disorient me.

The third snippet was taken when I am fifty feet away from the island and I was slowly walking closer.

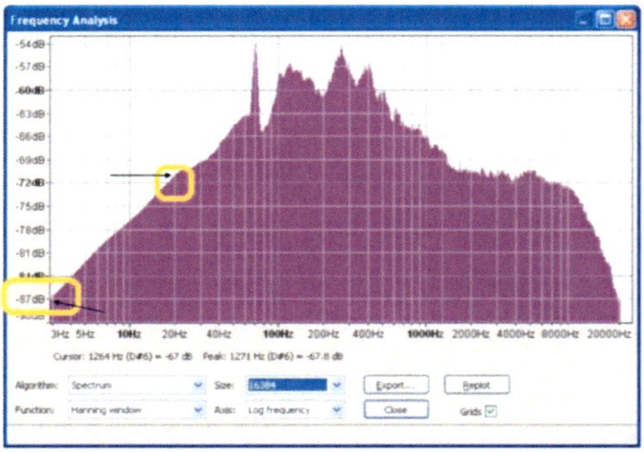

The infrasound amplitude averages have fallen slightly as I approach the Bigfoot hiding on the island. The 20Hz amplitude has dropped to an average of-73dB and the 3Hz has dropped to an average of -87dB, which is an 8dB drop for the 20Hz and a 4dB drop for the 3Hz.

The fourth snippet is the actual event. I am standing on the shoreline edge filming across to an overgrown island. The distance from where I am standing and where the Bigfoot is located on the island is approximately thirty feet. The water depth is only about three feet and covers an old road bed that runs in between the mainland and the island.

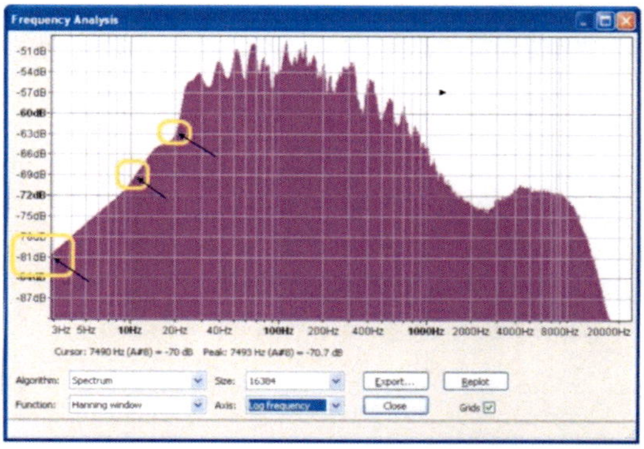

The entire range of infrasound frequencies has risen by an average of 10dB. 20Hz frequency is now averaging amplitude of 63dB an increase of 10dB. 3Hz average amplitude has increased by 7dB to 80dB. The real time analysis using the "Wavepad" software shows the highest spike during this snippet was -33dB for 20hz and -35dB for the 10hz frequency. The real time analysis also shows a variation in the amplitude of the infrasound frequencies during the snippet. These variations in my opinion demonstrate that the infrasound is originating from a biological source not a manmade source. You can observe what appears to be the Bigfoot taking breaths during the snippet. The infrasound frequencies drop off the scale for approximately half a second then jump back up to the higher amplitudes.

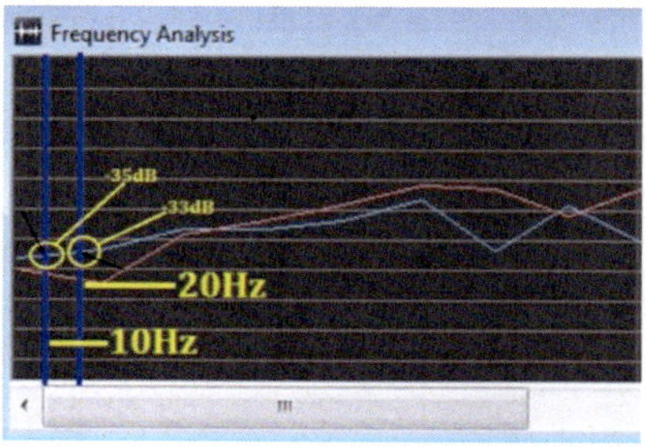

The above graph shows the spike in the snippet for the 10Hz and 20Hz frequencies.

Though I cannot hear it I think it is definitely having an influence on me. It is well documented that infrasound can have a range of negative effects on a person including nausea, causing a feeling of dread or panic, memory loss, emotional distress, hallucinations, blurred vision, and even a loss of consciousness.

The Department of Health and Human Services released a study, “Infrasound - Brief Review of Toxicological Literature”, that chronicles these affects and more.

(http://ntp.niehs.nih.gov/ntp/htdocs/Chem_Background/ExSumPdf/Infrasound.pd)

In the fifth snippet I am still standing in the same location but I have now turned my back on the Bigfoot and I am starting to walk away slowly.

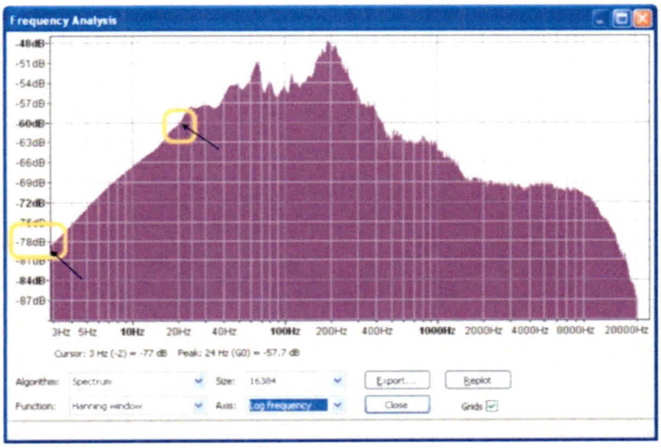

The infrasound range is now at its strongest amplitude of the entire event. I have just turned my back to the Bigfoot and have started to walk slowly away when the Bigfoot increases the infrasound intensity. The 20Hz frequency is now averaging -60dB and the 3Hz is averaging -79dB. The entire infrasound frequency range's amplitude has increased by an average of 20dB from the control snippet.

The Wavepad analysis shows the spikes in the amplitudes during this snippet climb to an astounding -40dB for 10Hz and -34dB for 20Hz. On this scale these amplitudes are the equivalent of someone screaming.

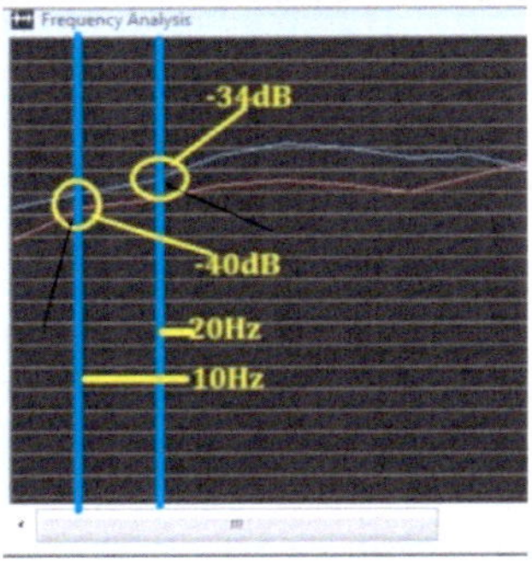

The effect on me was a loss of memory and a lack of reaction to what I was seeing. I have no recollection of the events that transpired on the video footage I captured. I do not seem to have any physical distress on the video. I show no signs of incoherence or loss of balance. I just simply turn and walk away as if I am under the Bigfoot's influence. I would compare this to the "Old Jedi mind trick" in the movie Star Wars where the Jedi would use mind control to suggest a person take a course of action not of their own free will.

In the sixth snippet I am approximately a hundred and fifty feet away from the island and the location of the event.

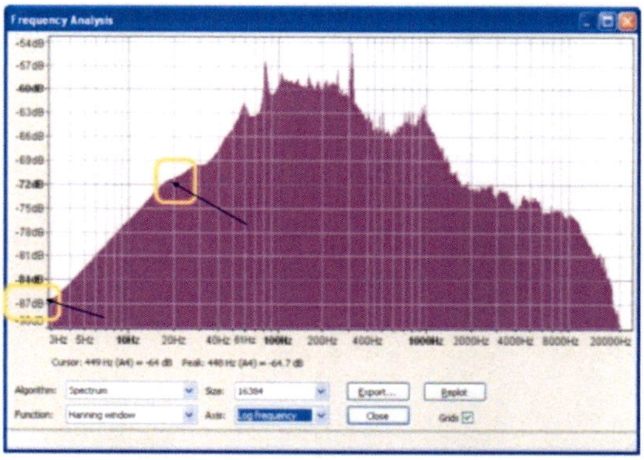

The infrasound range's amplitude is beginning to become less intense as I put distance between me and the Bigfoot. 20Hz is at 73db a drop of 13dBs and the 3Hz has dropped to -87dB which is a reduction of amplitude by 12dB. I still have not mentioned the encounter or had any reaction in this point of the video. I continue on doing my business in the research area as if nothing has occurred.
In the seventh and final snippet I am now approximately a mile away from the island and the location of the event.

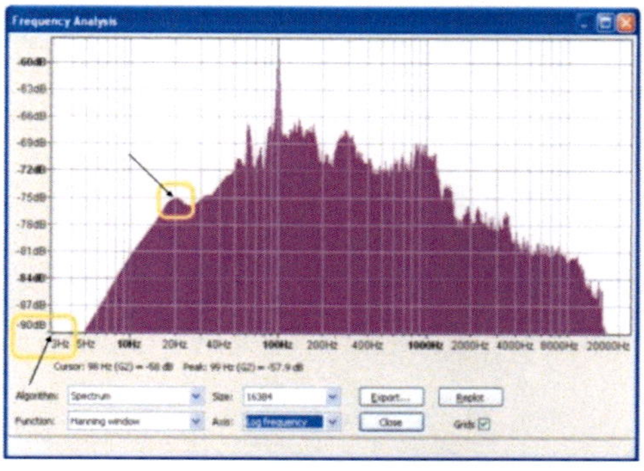

The entire infrasound frequency range has declined to negligible levels. 20Hz is at an average of -75dB and 3Hz to 5Hz is off the scale below -90dB.

Conclusions:

The following chart graphs the frequencies and the amplitudes before, during, and after the event. This demonstrates the infrasound activity and the spike in the amplitudes during the event (10 Ft Sneering and 10 Ft Back Turned).

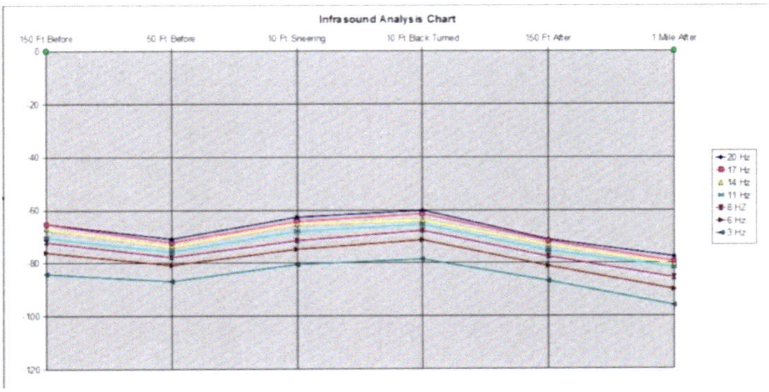

I am not an acoustic expert or a scientist. My findings are based on observation and common sense. I think that I was under the influence of infrasound during my encounter with the Bigfoot on April 30th, 2010. The Bigfoot manipulated my perception and sanitized my memory. Even more disturbing was the fact that I did not react to observing the Bigfoot. I had to have initially recognized what it was and where it was hiding. I made two attempts to zoom in on the Bigfoot and get a close up video. Sometime during this process I was subjected to the influence of infrasound and strongly influenced or "brain washed" into walking off. It is almost like my memory was wiped clean and I was given instructions to leave and I did.

I think the most interesting aspect of the event is my exposure to a combination of frequencies from 3Hz to 20Hz at different amplitudes and not just one particular frequency. I have read studies that claim 20Hz can cause hallucinations and that 7Hz is the resonant frequency of the body's organs and can cause internal organ damage or death. I think it logical to conclude that the combination of different frequencies/amplitudes at the same time could be the key to causing the affects I experienced. The mixing of the frequencies and at the correct intensity is something the Bigfoot has mastered and can take advantage of at will.

The ramifications of this conclusion are far reaching. If the Bigfoot have this ability and can produce infrasound without limitation it will be difficult for a lone investigator to capture video evidence of a Bigfoot. If an investigator successfully finds a Bigfoot all it has to do is subject the investigator to infrasound and simply walk away leaving the investigator unaware he/she encountered a Bigfoot. This is why I think it is very essential that the investigator never turn their video camera off in the woods. I also think investigators should have a rear facing camera recording at all times.

More study and investigation of the infrasound ability of the Bigfoot is necessary in order to determine the range and effectiveness of this ability. Is it directional in line of sight or effective in all directions if one is close enough? Is infrasound being used as a communication method as well much like elephants and whales use it? Finally the most important question is can we learn to detect it and use this to our advantage to locate the Bigfoot?

In closing I think the evidence I have presented makes a strong case that:

- The Bigfoot can produce and use infrasound.
- That I was under this influence and was profoundly affected by it during this encounter.

This chapter may be refuted by acoustical experts or scientists in this field of acoustics. I have openly admitted I have no formal training in the discipline of acoustics or infrasound. If I am refuted it would actually be a good thing. Currently it is virtually impossible to get scientist to look into this phenomenon. If this chapter causes "experts" to look at the Bigfoot and infrasound then it will be a victory for the Bigfoot research community.

Source Material:

A Case for Infrasound - Cliff Barackman
http://www.northamericanbigfoot.com/MyResearchPages/articles/CliffArticles/ACaseforInfrasound.html

Infrasound – Wikipedia
http://en.wikipedia.org/wiki/Infrasound

Acoustic Trauma: Bioeffects of Sound - Alex Davies BigfootA Honours
"Many of the most profound effects of sound are attributed to infrasound in the region of 7Hz. This corresponds with the median alpha-rhythm a frequency of the brain. It is also commonly alleged that this is the resonant frequency of the body's organs and hence organ rupture and death can occur at high intensity exposures."

http://schizophonia.com/installation/trauma/trauma_thesis/index.htm

Chapter 47 – Theories – The Bigfoot have the ability to cloak or camouflage itself using acoustic.

The "Predator" Effect

Background: Over the last two years of my research I have noticed a phenomenon that I occasionally will capture on video recordings. I will usually get the gut "feeling" I am being watched. I then stop and take video of the area that where I think the Bigfoot is concealed. I will pause and video for a minute or two and sometimes zoom in and out of the area then move on. When I review the video at a later date I will notice a "blurry" area in the video. The foliage in front of and behind this "blurry" area will be in crisp focus. In a small number of these "blurry" areas a faint figure can sometimes be seen. The figure is always out of focus but recognizable features can sometimes be seen in the "blurry" area such as a nose, mouth, eyes, arms, shoulder, and upper trunk. I have checked my video camera and it is functioning correctly. The lens is clear of debris or finger smudges. The "blurry" area is in different locations on the frame, left, right, and middle. I am not seeing this "blurry" effect with the naked eye while I am in the field. I only seem to capture it on the video camera.

I began to do research on what could be causing this or what it could be. I found many eye witness accounts where the person reports seeing a Bigfoot then it just "disappears into thin air". Other witnesses have reported seeing a figure that resembles the alien in the "Predator" movies when using the artificial "cloak". Native Americans have long believed the Bigfoot has supernatural powers to become invisible or shape shift into other animals. The Native Americans believe this explains how when seemingly cornered with no route of escape the Bigfoot can just disappear from sight.

Theory: The Bigfoot has the ability to use acoustics (infrasound) to excite and control the water vapor molecules in the air around them. They use the water vapor as a prism to reflect and diffuse the light away from their bodies creating a cloaking affect. The Bigfoot will appear to blend in with the environment around them to create a natural camouflage. This camouflage effect is created because the eye is seeing the objects that are around the Bigfoot at different angles.

This is the same concept that is used in a commercial product called "GhostBlind". (http://www.youtube.com/watch?v=U2EeappH-zY) This product uses a mirror to reflect back to the viewer the surrounding environment be it rocks, green vegetation, dead leaves, trees, etc. Instead of using a mirror the Bigfoot are using water vapors in the atmosphere.

In the following example please note how the edges and angles give the blind the "Predator" affect.

Supporting Documentation:
Date: May 22nd, 2010 - Location Research Area B, National Park

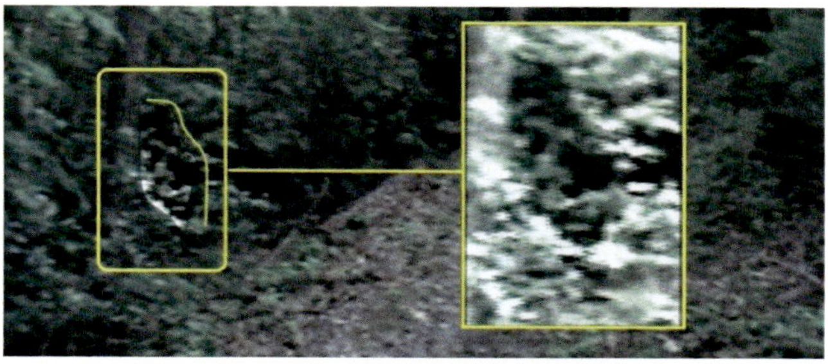

The Bigfoot is green in color and blurry. I propose that camera is able to capture the image where I could not see the Bigfoot with my naked eye.

05/22/2010 06/23/2011

The above photograph on the left shows the area during the Bigfoot sighting and after. The side by side photographs on the right show a close up of the Bigfoot in stealth mode in color and black and white for more detail.

Date: June 19th, 2010 - Location: Research Area B, National Park

Please note how blurry the subject is compared to the surrounding trees and vegetation. Both trees in the foreground and back ground are in clear focus. This is an example of the "blurring" effect I capture on my videos intermittently. Also take notice that the sun is shining bright on the limbs and vegetation directly in front and beside of the "blurry subject". Is the area dark because the sunlight is being reflected off by water vapor at different angles? Please also note that I am not filming this area randomly. My attention was drawn to this area by a "gut" feeling I was being watched. I paused in this area for approximately thirty seconds. I even zoomed into the thick foliage in the center of the above still capture. I could not see the "blurry" subject with the naked eye.

After approximately twenty seconds I zoom in to the center of the vegetation to the upper right of the subject then zoom back out to normal size. The above still capture is at approximately 30 seconds. Note the blurring around the subject has diminished significantly. The subject is still out of focus but now resembles the still captures from May 22nd, 2010. I have a theory that the Bigfoot was running out of breath and was losing the ability to maintain the amplitude needed to keep the water vapor "tuned" and reflecting light away from its body. Also note that the sun light is not as bright because it was hidden by a passing cloud during this point in the video. Did the low light conditions also affect the Bigfoot's ability to maintain the cloak? Immediately after the above frame was captured I turned and continued down the trail.

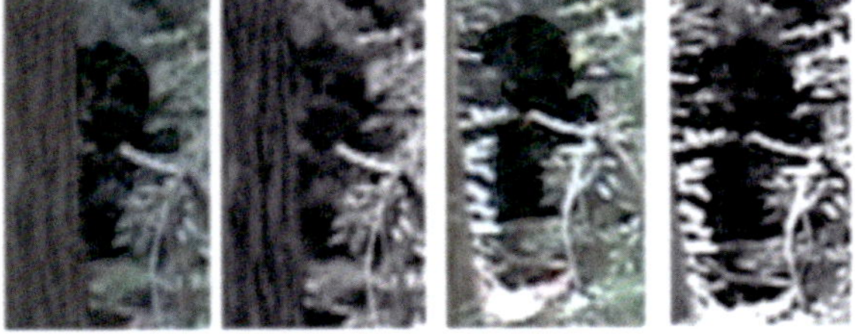

In the above photographs I am comparing how the mirror blind distorts the surroundings as it reflects them and how the Bigfoot is distorting its surroundings as they are reflected. The main difference is the size and number of the "mirrors". The blind only has six large panels but the Bigfoot is using millions of water molecules to refract the light.

Date: July 11th, 2010 - Location: Research Area A

I am walking down the horse trail as I approach two Bigfoot in complete stealth mode on the left side of the trail. The Bigfoot in the photograph below is standing up and appears transparent.

To the right of the transparent Bigfoot is another one in stealth mode. It is approximately three feet off the trail. It is literally in arms reach. The photographs show three successive frames where the Bigfoot is a light green blur as I approach in frames 1 and 2. Then by frame 3 he is becoming visible.

Frame 1

Frame 2

Frame 3

The Bigfoot comes into focus in Frame 3. If you look close you can see its white teeth, eye, and brow.

I am filming as I am walking so the camera is moving back and forth and up and down. When I get almost beside the Bigfoot it becomes visible to the camera. Remember, I do not see it as I am walking down the trail. I move closer to the Bigfoot and the facial detail becomes more visible.

An interesting side note about this encounter is that I pause right beside the Bigfoot and mention on camera that I feel one very close but I look the opposite direction! I turn my back on both of them. I film the wrong direction for approximately two minutes then move on down the trail never turning around. I wonder if they were influencing me with infrasound and "encouraged me" to move on without turning around?

Date: June 5th, 2010 - Location: Research Area B, National Park

I capture just a few frames of a Bigfoot hiding in the middle of heavy vegetation. It is only approximately one hundred feet from the parking area of the trail head. I had the feeling I was being watched and should have stopped and filmed the area thoroughly. Unfortunately I was in a hurry to get to other areas so I pointed the camera in the direction I thought the Bigfoot was hiding and filmed as I walked. This was a mistake. The Bigfoot was no more than fifty feet off the trail. He is in total stealth mode. I have debated if the algorithm of the digital camera turned him green or he was in stealth mode. I am now convinced he was in stealth mode. He was green because that is the color of the vegetation around his body.

The above composite shows the Bigfoot in "Stealth Mode". Note that his face is green and matches the foliage around him. The insert photograph on the left has been converted to black and white. This allows you to see the facial detail.

Conclusions:

The theory I have proposed could be categorized as "Science Fiction". Scientists have only begun to investigate acoustics and how it affects us and the world around us. The science of acoustics is just now in its infancy. Early experiments are showing that sound waves can do amazing things to include levitating objects, boiling water, crumble solid rock, and have numerous physical and physiological effects on humans.

Is the use of acoustics a skill the Bigfoot has mastered? Do they use it on a regular basis to communicate and protect themselves? Is this skill taught to their young and passed on from generation to generation?

The theory I have proposed could be just one possible answer to these questions. I have only provided anecdotal evidence at best. It will take field time, extensive video documentation, and some help from sympathetic experts in the acoustics field to gather the necessary evidence to prove or disprove my theory. I do however think it is plausible based on my observations and research. It is a good starting point to build on.

Research Area A

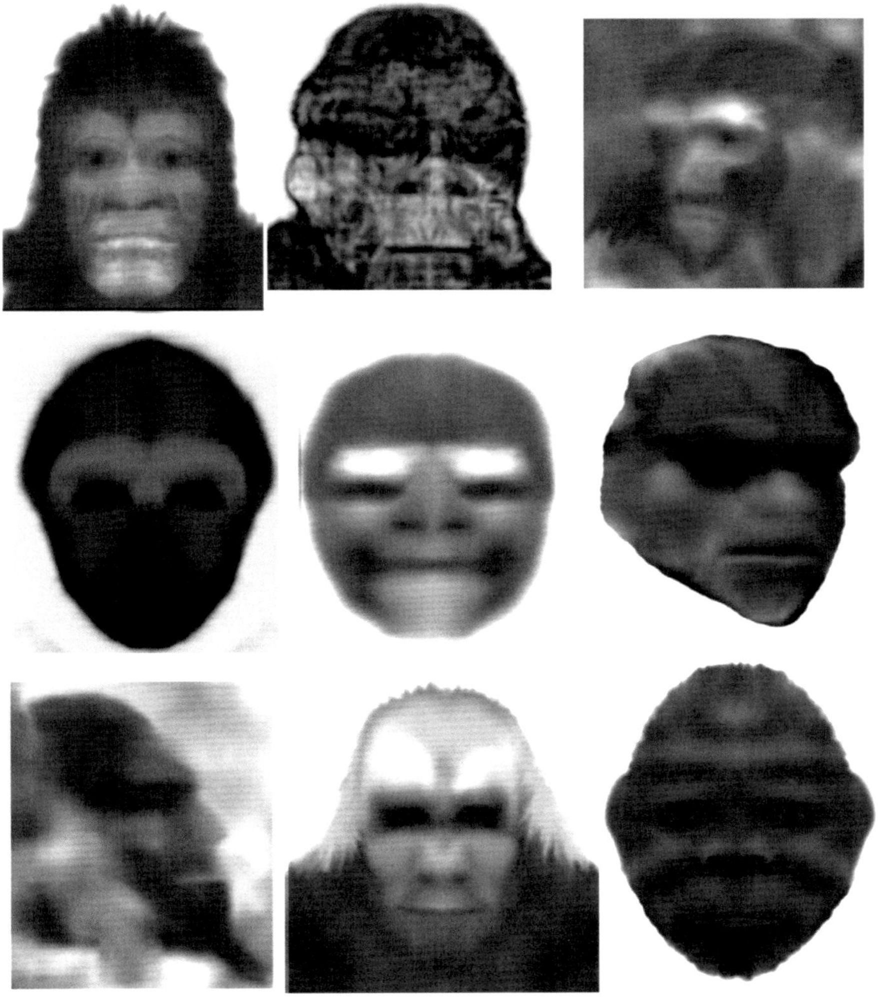

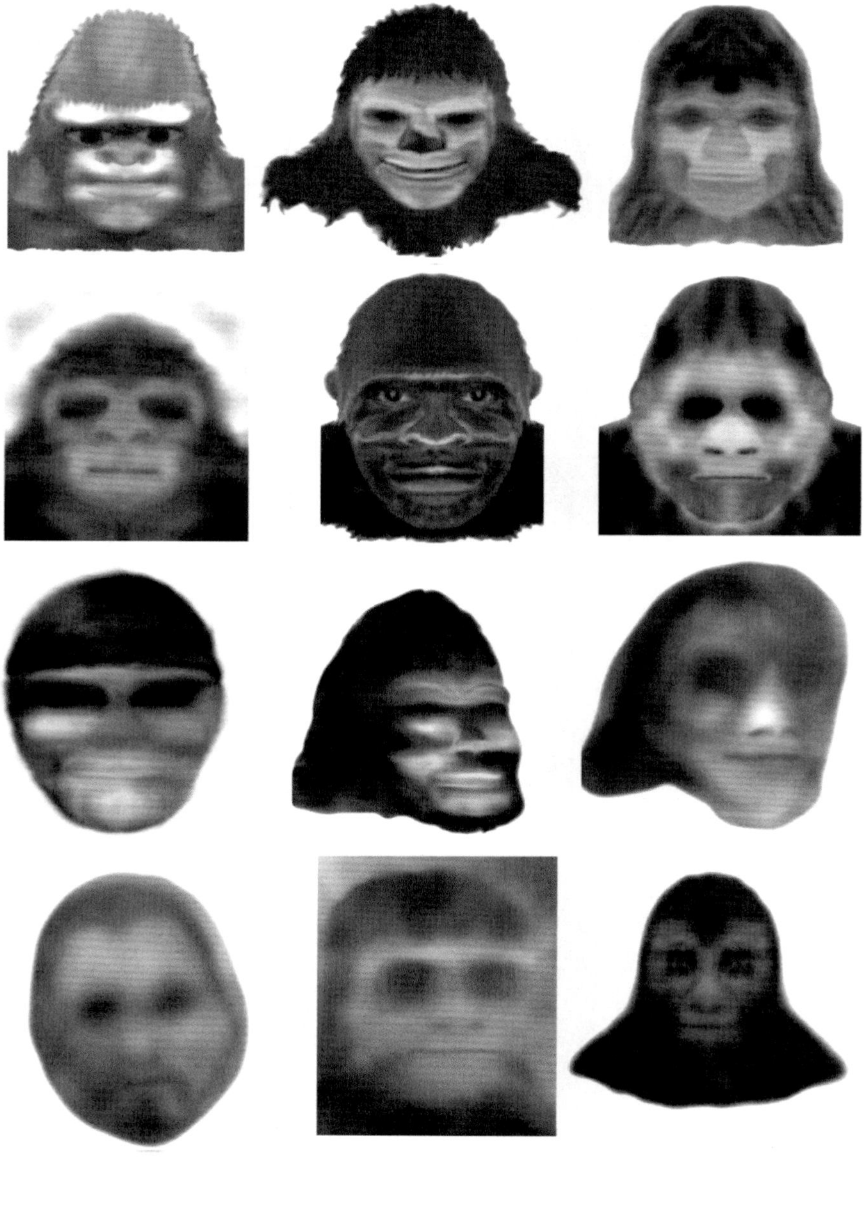

Research Area B – National Park

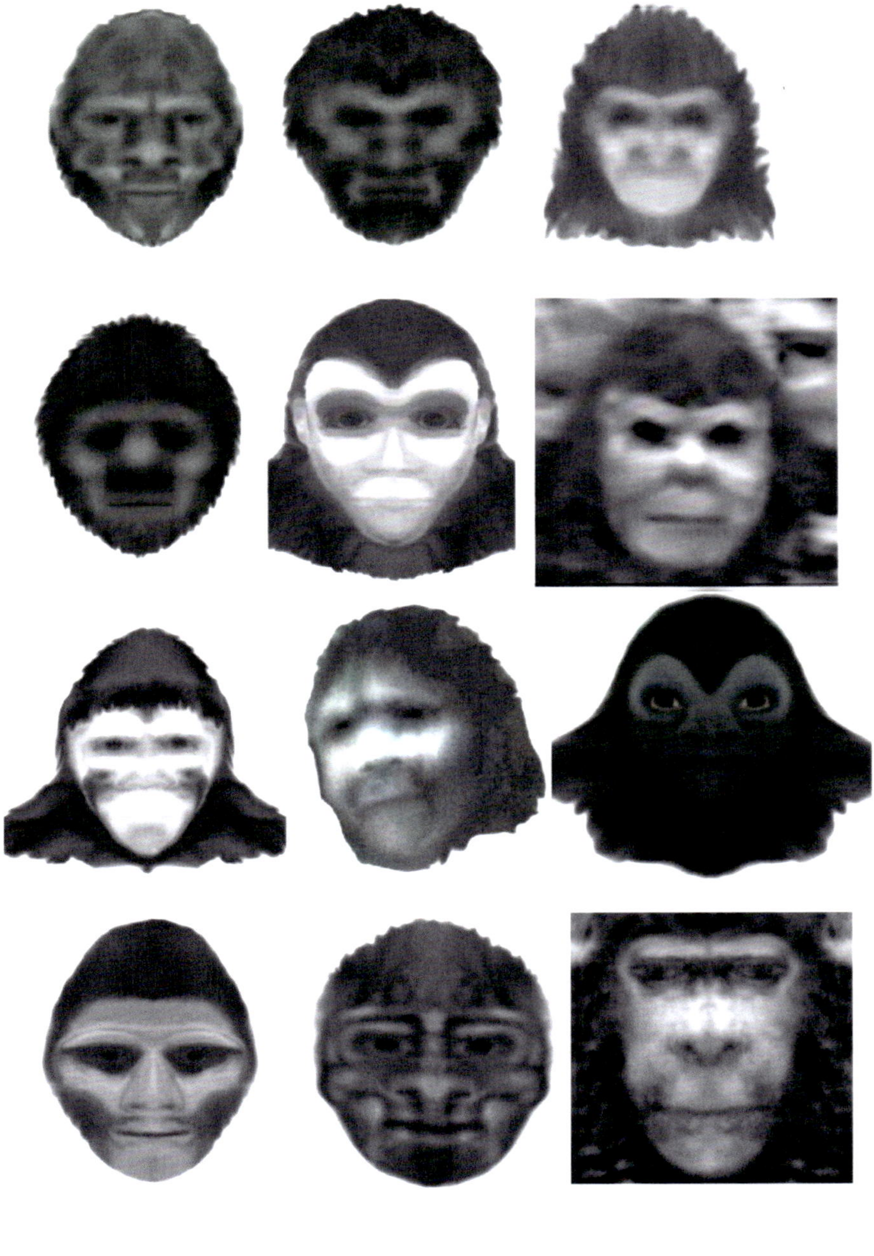

Chapter 49 - The DNA Project (aka Ketchum DNA Study)

In June of 2010 I began to correspond with David Paulides the Executive Director of North American Bigfoot Search (NABS). I realized that I needed assistance with my research. I was encountering phenomenon and having experiences that were extraordinary. I needed to find a support network that I could partner with in my research. I began researching the different Bigfoot Research organizations and I found NABS to be the most professional, respectful and responsive group. I had read David Paulides' book "The Hoopa Project" and was already impressed with his logical approach to Bigfoot research, so I introduced myself to the group via email correspondence.

I began to bounce questions off NABS and relate my experiences to them on a regular basis. I also began to collect hair samples and would forward them to NABS periodically. I also sent NABS videos and photographs of the Bigfoot I was encountering in my research areas. The relationship matured and in February of 2011 David invited me to join the NABS Team as a Field Researcher. I gladly accepted his offer.
http://www.nabigfootsearch.com/nabs_team_members.html

NABS had already partnered with Dr. Melba Ketchum and had begun a ground breaking project to collect Bigfoot DNA specimens from all over the North American continent. You can read in detail the studies scope and background at NABS website –
http://www.nabigfootsearch.com/bigfoot_dna.html.

I had begun working with NABS on this project in the spring/summer of 2010 even before I became a team member. David and I discussed the matter and the need to humanely collect as much DNA from the Bigfoot as possible. Through collaboration we came up with an innovative strategy. I took clear packing tape and placed it sticky side out. When the Bigfoot would lean or rub up against the tree their hair would stick to the tape and be pulled out by the roots. The root of the hair contains the DNA. I placed this "hair trap" around the tree the Bigfoot where hiding behind when I captured them on the trail camera in August of 2010. The trail cameras were still deployed in this area on the same trees. I was hopeful the Bigfoot would hide behind this tree again while observing the trail camera. The effort was successful and I collected a large amount of Bigfoot hair.

I stayed out of the area for almost a month. When I return I find the tape has been torn off the tree and was lying on the ground approximately fifteen feet away from the tree. When I look at the tape I find I was successful! The tape contains dozens of black hairs with their roots or skin tags attached.

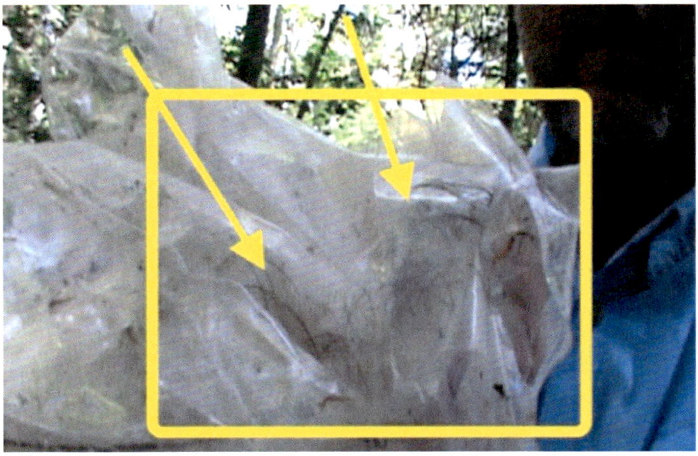

The above photograph shows the tape and the hair that was pulled out of the Bigfoot right after I found it on the ground. I placed the tape with the hair in a plastic bag and placed rubber gloves on my hands when I handled the tape to avoid any contamination.

On this same day I also found a second hair sample in the broken end of a small holly bush. I use rubber gloves to collect the hair sample and place it into a plastic bag. Once I returned home I removed the section of tape that contained the hair. I also sorted out the hairs taken from the holly bush. Using rubber gloves and sterilized tweezers I packaged the hair samples separately and sent them to Dr. Ketchum's laboratory in Texas.

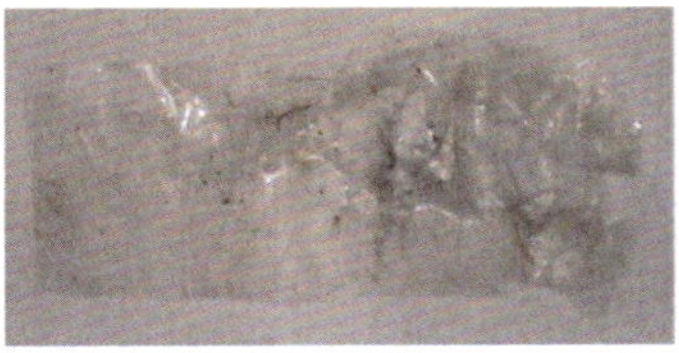 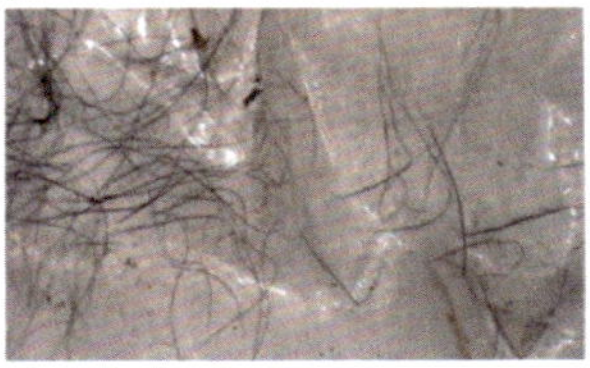

The above photograph on the left shows the section of tape that contained the Bigfoot hair. The photograph on the right is a close up of the hairs in the tape.

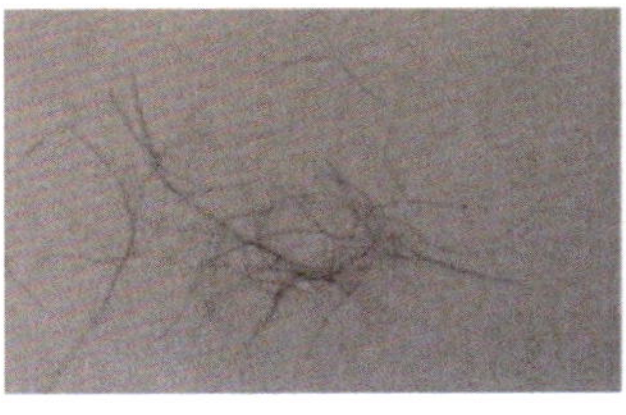 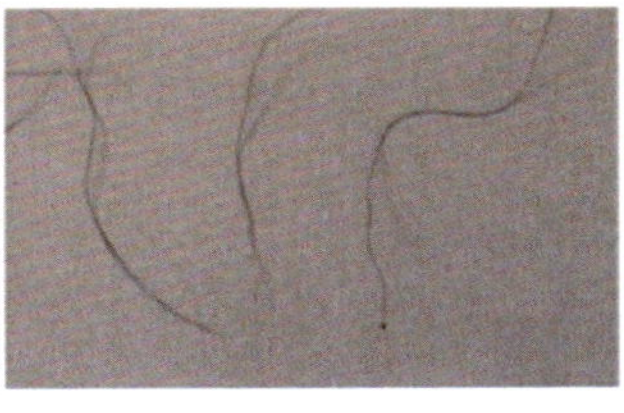

The above photograph on the left shows the hair collected from the holly bush. The photograph on the right shows an example of hair with the roots or skin tags attached.

I also noticed when I was untangling the tape that contained the Bigfoot hair there was a large finger print in the tape. I had been very careful not to touch the tape when I put it around the tree. This is a Bigfoot fingerprint. It is approximately two times larger than a human's finger.

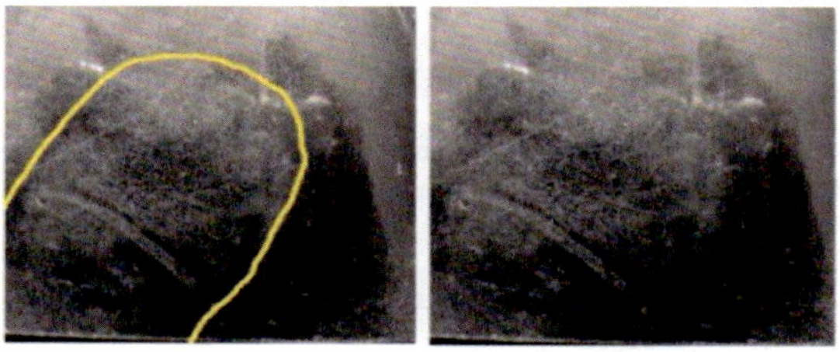

Perfecting the "Hair Trap"

Trail cameras were not an effective long term solution for baiting the hair traps. I realized I needed to bait my hair traps in such a way that I could get the Bigfoot to press against the exposed adhesive side of the packing tape. I knew that the Bigfoot had a taste for bacon. So I decided to take bacon grease and smear it into the bark of a tree approximately eight feet off the ground. Then at approximately six feet I would wrap packing tape around the tree adhesive side outward. When the Bigfoot would lean against the tree to lick the bacon grease off the bark its hair would get caught in the tape and pulled out with the root attached.

The National Park had a large population of Black Bear so this method would not work in that area, but Research Area A had no bears near the area. I would set my new and improved hair trap there.

The above shows a typical example of a hair trap. I have bacon grease smeared onto the bark of the tree.

Below the bacon grease I have wrapped packing tape around the tree adhesive side out. I learned through trial and error that placing three sections of tape around the tree was the most productive. I have collected many hair samples this way. I know the Bigfoot can see the tape but since it is only a minor discomfort for a moment they endure the pain for the bacon grease.

I also use another method of collecting hair samples that I learned from the holly bush. I go along known Bigfoot walkways and break limbs off so that the jagged end is facing toward the trail. I break them off around four feet high. I want the breaks below eye level where the Bigfoot will not notice then and hopefully rub its leg against them pulling out some hairs by the root.
This method requires no bait and has also proven successful. I make these "natural" hair traps in the National Park since the bears would interfere with a hair trap baited with food. This method has also been effective in producing results.

I continue to collect hair samples via the hair traps but I also have found hair by just being observant. It appears that the Bigfoot may shed their hair in the spring when the weather warms. In the spring of 2011 I found a large amount of hair on the ground near a heavily traveled Bigfoot route. The hair was sent to Dr. Ketchum for testing. I have collected hair samples from multiple locations, Research area A, Research Area B (National Park) and my backyard. All were sent into the Ketchum DNA Study.

The DNA Donors

As I mentioned previously I am involved in what is known as the "Bigfoot DNA Project" or the "Melba Ketchum DNA Study". Over the span of almost three years I submitted numerous hair samples and candy wrappers that I suspected contained saliva. The DNA Study was officially published on February 13th, 2013. The study has been electronically published at the Denovo Journal Website. (**Novel North American Hominins -Next Generation Sequencing of Three Whole Genomes and Associated Studies**)

The official website for the study: http://sasquatchgenomeproject.org/

I submitted numerous samples to the study of hair and saliva. Eleven (11) hair samples passed initial review as being possible Bigfoot hair. The below is an excerpt from the study "Materials and Methods S1" (https://sasquatchgenomeproject.org/groups-and-submitters/)

Hair Analysis:
Hair samples were sent to the Southwestern Institute of Forensic Sciences (Dallas, TX) for analysis. The samples were evaluated visually, stereoscopically, and by light microscopy to determine human or animal origin. Hairs that were classified as potential novel hominid were also evaluated for DNA typing potential by examining for root material.

Only hairs that were not human in appearance and could not be identified as any other species were utilized in this study.

The samples were processed and the mitochondrial DNA extracted and sequenced. The 11 samples that were chosen from the samples I submitted were determined to be Sasquatch (Bigfoot). The sample numbers from Table 1 of the Study are 2, 3, 11, 38, 42, 70, 82, 122, 125, 126, and 127. From this group of samples three had enough genetic material to run a full mitochondrial DNA sequence. These samples numbers were 2, 11, and 38. It was determined that the female linage of the Bigfoot was human and the male linage was from a still unknown hominin.

During the collection of hair samples I was able to capture a nighttime infrared video of four Bigfoot taking turns peeking at my trail camera. I subsequently put a hair trap consisting of packing tape adhesive side outward around the tree that the Bigfoot had used to hide behind. When I returned at a later date I found tape had been torn off the tree and contained a large amount of hair that was extracted with the root or skin tags still intact. This would become sample number 38 in the DNA Study.

I mentioned earlier in this chapter that I noticed four Bigfoot taking turns peeking at the trail camera from between two trees. In the trail camera video I identified four different individual Bigfoot. It appeared they were taking turns looking between the two trees at the camera. The height of the individuals ranged between seven and eight feet. In one instance it appeared that one subject jumped onto the shoulders of the Bigfoot in front.

One of the following Bigfoot is the DNA donor of sample number 38.

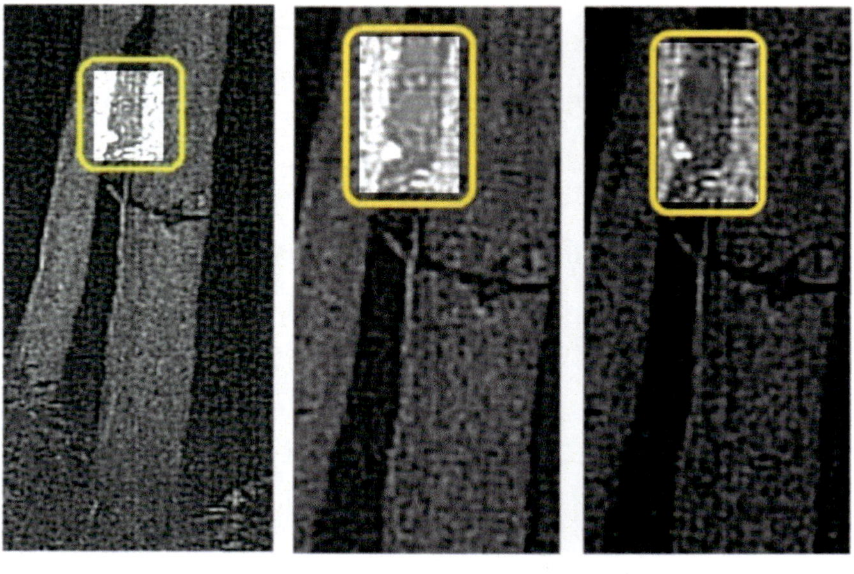

Close ups and sketches of each Bigfoot.

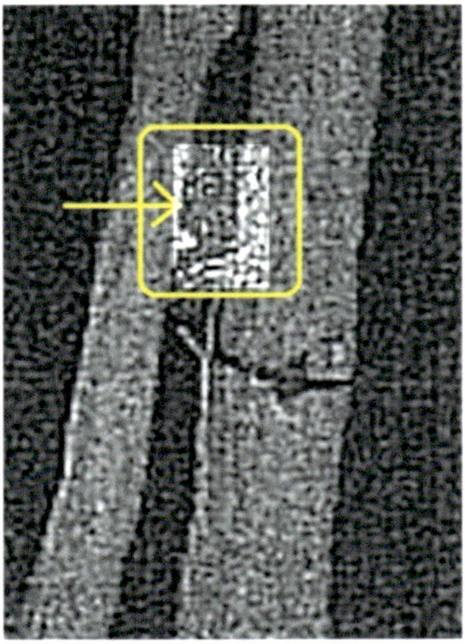

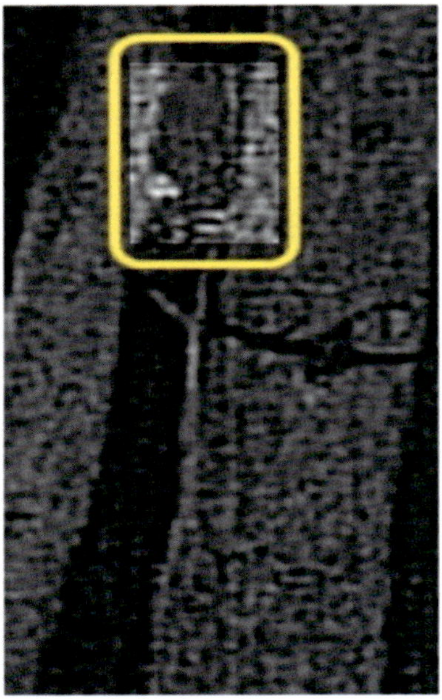

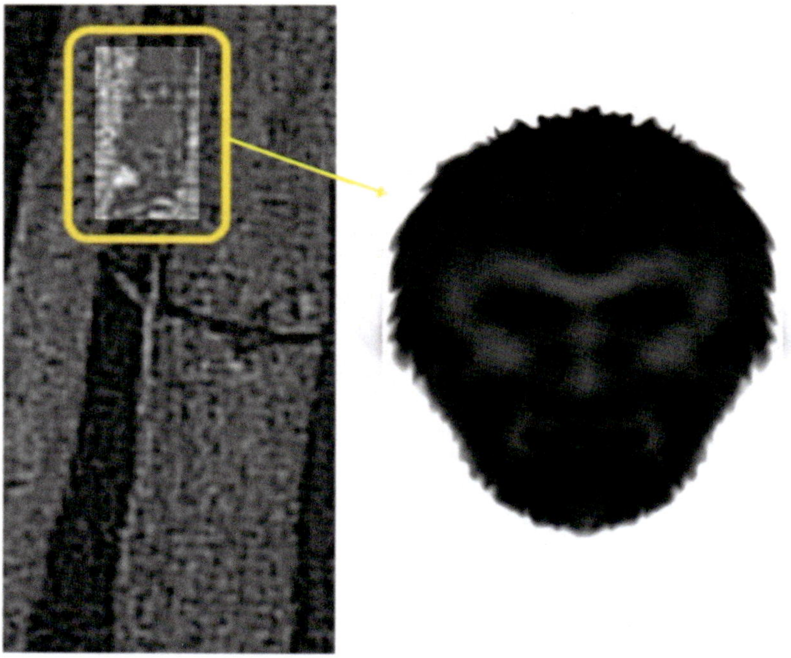

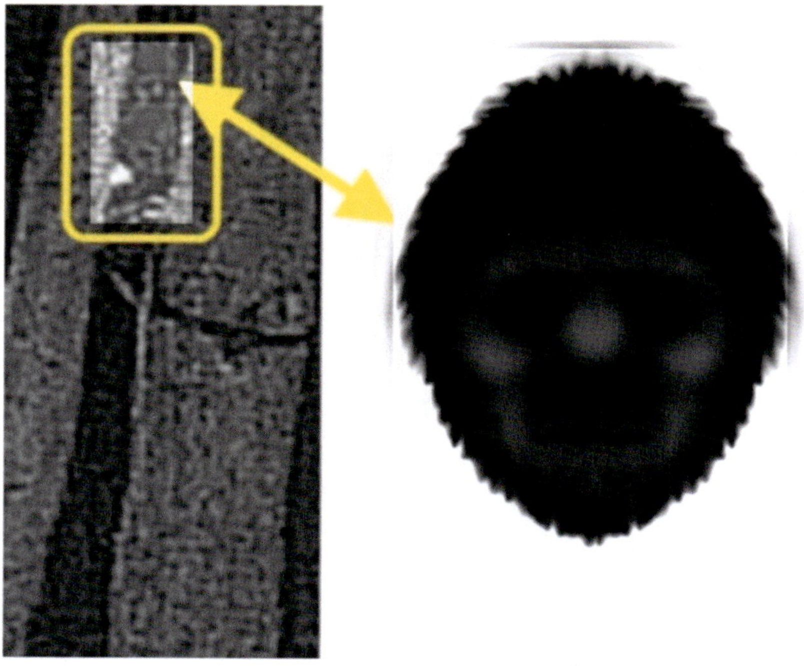

The probable hair donor for sample 96 or 97 that came from the feeding station at Research Area A. (Public Land).

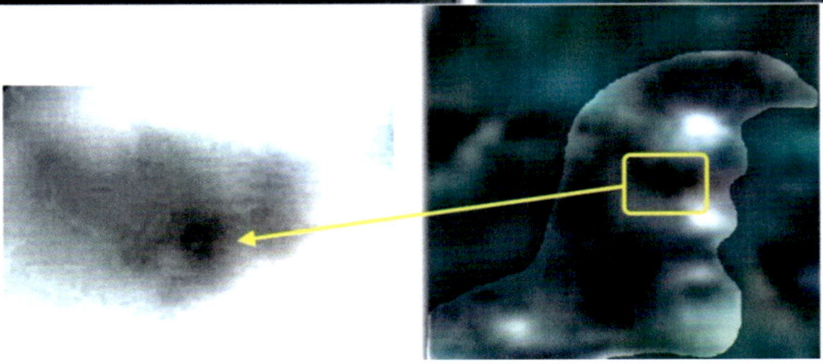

The Ketchum DNA Study – Press Release

Researchers Sequence Sasquatch Genome, Novel Hominins Extant in North America Next-Generation Whole Genome Sequencing Proves Extant Human Relative, Findings Published in Denovo Journal of Science

A team of eleven scientists with expertise in genetics, forensics, pathology, biochemistry, and biophysics has sequenced three whole nuclear genomes from a novel, contemporary species of hominin in North America. The study, "Novel North American Hominins, Next Generation Sequencing of Three Whole Genomes and Associated Studies," which analyzed DNA from a total of 111 high-quality samples submitted from across the continent, appears in the inaugural issue of Denovo: Journal of Science (http://www.denovojournal.com) on February 13.

The team, led by Dr. Melba S. Ketchum, DVM, of DNA Diagnostics in Nacogdoches, TX, sequenced the three whole nuclear genomes using the next-generation Illumina HiSeq 2000 platform at University of Texas, Southwestern from a tissue sample, a saliva sample, and a blood sample. The three genomes all attained Q30 quality scores above 88 on the Illumina platform, significantly higher than the platform average of 85, indicating highly-purified, single-source DNA with no contamination for each sample. The three Sasquatch genomes align well with one-another and show substantial homology to primate sequences.

In addition to the three nuclear genomes, Ketchum's team also sequenced 20 whole and 10 partial mitochondrial genomes from the Sasquatch samples. In contrast to the nuDNA results, the Sasquatch mtDNA was fully modern Homo sapiens, indicating that the species is a hybrid cross between modern Homo sapiens in the maternal lineage and an unknown hominin male progenitor.

Specimens yielding DNA were obtained, purportedly from elusive hominins in North America called Sasquatch. Sequencing and genotyping were performed in addition to histopathologic and electron microscopic examination of a large tissue sample.
Mitochondrial whole genomes were consistent with modern humans. In contrast, novel data were obtained when nuclear DNA was sequenced. Next generation whole genome sequencing was performed on three samples. Phylogeny trees generated showed homology to human chromosome 11 and to primate sequences. The data indicates that the Sasquatch has human mitochondrial DNA but possesses nuclear DNA that is a structural mosaic consisting of human and novel non-human DNA.
"Novel North American Hominins, Next Generation Sequencing of Three Whole Genomes and Associated Studies."
Authors: Ketchum MS, Wojtkiewicz PW, Watts AB, Spence DW, Holzenburg AK, Toler DG, Prychitko TM, Zhang F, Bollinger S, Shoulders R, Smith R.

North American Bigfoot Search – Press Release

Dr. Ketchum and her team of scientists and laboratories have been able to sequence three complete genomes, getting identical results independently verified by different labs. This is not a casual everyday finding, this will cause the scientific community to stand up and take notice. Many people have claimed to know what our results were, nearly everyone was wrong, even some high profile media people that have just recently made comments, all were wrong.

Dr. Ketchum originally found the combination to unlock Bigfoot DNA and utilized top scientists in various fields to validate her results. The results were independently verified with the group silently sitting on these findings for months, as the results were validated a third and fourth time. Remember, this isn't one total genome sequence; it's three with dozens and dozens of other findings validating the results. The exact number of specimens, the number of mitochondrial results and the specifics of the scientific white paper will be released very, very soon. The paper is very complex and filled with amazing findings. Before the world is enveloped in Bigfoot, Sasquatch mania and we lose sight of the real heroes of this project, I need to address a few of the VERY best researchers in the world. **Scott Carpenter, Richard Hucklebridge, Rob Alley and Harvey Pratt** have contributed to the knowledge base of the scientists in a way never before accomplished. It was the vision and professional manner that these fine men have and the ability to overcome public ridicule and outright hostility that changed the course of the understanding of Bigfoot forever. Where the majority of Bigfoot researchers failed in getting specimens on multiple occasions utilizing innovative methods, these men did it.

The one person that was able to unlock the safe to understanding Bigfoot DNA, Dr. Melba Ketchum. Where every other scientist gave up and claimed contamination, Dr. Ketchum continued to move forward. Where every other scientist was unable to turn the corner, Dr. Ketchum did. This lady did not understand the word "defeat," she is a true pioneer in her field!

We need to express a sincere "Thank You" to the dozens of others who politely and professionally contributed to our project and the pool of specimens. We could not have done this without you!! Your patience and adherence to the Non-Disclosure Agreement shows your belief in the project and your stoic professionalism. The days ahead are going to be exciting.

David Paulides
Director
North America Bigfoot Search

Chapter 50 - Closing Thoughts – "Monsters" in the Woods?

The Bigfoot Field Journal Volume 1 just scratches the surface of what is going on in my research. I found that there are many things that are unknown to the modern world. I have been forced to open my mind to new ideas and new realities.

The Bigfoot is not the only cryptid roaming the woods of North American. I inadvertently found another creature while researching the Bigfoot or it would be more accurate to say this creature found me. In volume two I will chronicle my encounters with the mysterious Dog Man.

I had only heard of this creature and did not think I would ever encounter one in East Tennessee. I thought it was probably a Bigfoot with a genetic mutation, but when I encountered one I knew it was something very different. The Bigfoot is human, but the Dog Man is something very different.

It all started on March 27th, 2011. I was in the public research area doing my thing. I had visited the stick shelter and worked my way to the island crossing. I was looking across to the island and filming. I also had the "back trail" camera on my shoulder filming behind me. Unknown to me at the time a dark figure was standing in the vegetation beside the thick and overgrown trail. The figure was dark brown-black and approximately six feet tall. I did not know this "thing" was behind me until I reviewed my back trail footage the week after my research visit.

To my horror about five feet behind me was standing a creature of myth! These things just do not exist, this cannot be! I replayed the video over and over. My mind did not want to process what my eyes were seeing. There was no doubt it was real. It moved several times in reaction to me turning my shoulders and when I did turn around it ducked down! I finally had to admit to myself I had HD footage of a "Dogman" less than five feet behind me!

I stepped through the footage a frame at a time. The snout clearly visible, the nostrils moving and wet as it was breathing! The right eye was plainly visible and moving! It was watching me! It could have taken me out anytime it wanted too. I had no clue it was within arms or claws reach of me!

I showed the video to some trusted friends and respected Bigfoot researchers. The reactions were mixed. Some did not want to deal with it. Others were ecstatic and could not believe it. The most common comment I received was "Wow, if you had a close up of a Bigfoot like that you'd be rich! Too bad it's a Dogman".

The story gets even better because these Dogmen were not content to just stand behind me. On May 22nd, 2011 I was back in this area, firearm at the ready, as I was putting up some hair traps and setting out bacon in a jar. The bacon in a jar was an attempt to get fingerprints. I had the lid on tight but had cut holes in the lid so the smell of fresh bacon was heavy in the air.

I had the feeling I was being watched and it was not a good feeling, I felt threatened! I see something dark black about thirty feet in front of me so I take my video camera and zoom in on this black object. My heart starts pounding and I draw my weapon and chamber a round! The black object is a Dogman! It is looking at me from around a tree. It is in heavy vegetation so I am looking through my viewfinder on the camera zooming in and out trying to get it in focus and see it better. Then its ears move, like when a dog hears a sound and begins twisting its ears, I look up over the top of the camera and make direct eye contact with this thing, I get a sick bad feeling! I look back down at the camera trying to get a better shot of the Dogman when my brain finally kicks in and I literally say to myself, "What the heck are you doing!! Get out of here now, stupid!"

I slowly begin to back out, running attack scenarios through my head. "If he charges on all fours stay calm aim for the head and shoulders, stay calm, squeeze off the shots deliberately, if he is on two legs center mass". I was convinced I was going to be attacked! To my surprise I was able to back out of the area without incident. My wife had dropped me off on this day, needless to say I called her to come get me immediately and she found me already on the road out of the area, walking fast, gun still drawn, "white as a sheet" as she put it.

This is why I no longer go into that area alone and unarmed. That was the first time in my life since my military service I felt my life was in imminent danger!

I went back the following week and found that the Dogman had jumped up on the side of a tree so it could look at me above the thick cover. I found the bark had been torn off the tree and there were deep "claw" marks in the tree.

This was not my only encounter with the Dogman and in Volume 2 I will explore the subject further.

I continue with my research with the goal of establishing a relationship with the local Bigfoot clans that visit my home and that live in my research areas. Each trip brings me closer to this goal. Trust can only be built with time and with the Bigfoot it will be on their schedule not mine. I hope by the time I publish my second volume I can tell you about my face to face encounters with the Bigfoot. Only time will tell.

God Bless.

Scott

To keep up with my ongoing research and developments in the Bigfoot DNA Study please visit my blog at: http://bf-field-journal.blogspot.com/

My YouTube Channel:
http://www.youtube.com/user/joebblack1963?feature=mhee

The Sasquatch Genome Website: http://sasquatchgenomeproject.org/

Made in United States
Orlando, FL
05 November 2026

90229883R00155